10th International Conference on Advanced Manufacturing Technologies (ICAMaT 2019)

IOP Conference Series: Materials Science and Engineering
Volume 682

Bucharest, Romania
10 - 11 October 2019

ISBN: 978-1-7138-0934-0
ISSN: 1757-8981

Printed from e-media with permission by:

Curran Associates, Inc.
57 Morehouse Lane
Red Hook, NY 12571

proceedings.com

Some format issues inherent in the e-media version may also appear in this print version.

Institute of Physics
Dirac House, Temple Back
Bristol BS1 6BE UK

Phone: 44 1 17 929 7481
Fax: 44 1 17 920 0979

techtracking@iop.org

Additional copies of this publication are available from:

Curran Associates, Inc.
57 Morehouse Lane
Red Hook, NY 12571 USA
Phone: 845-758-0400
Fax: 845-758-2633
Email: curran@proceedings.com
Web: www.proceedings.com

TABLE OF CONTENTS

INTEGRATED PRODUCT DEVELOPMENT

FLEXIBLE AND INTELLIGENT MANUFACTURING SYSTEMS

IOP Conf. Series: Materials Science and Engineering **682** (2019) 011001 doi:10.1088/1757-899X/682/1/011001

The 10th **International Conference of Advanced Manufacturing Technologies, ICAMaT 2019**, was held in the University POLITEHNICA of Bucharest, Romania, Faculty of Industrial Engineering and Robotics, on 10th and 11th of October 2019, with the remarkable participation of experts from Poland, Serbia and Romania.

The first ICAMaT Conference was held in 1999 under the acronym CITAF (derived from the abbreviation in Romanian) at the initiative of the Manufacturing Engineering Department (TCM) and had a national character. The following Conference sessions were held almost every two years and attracted increasingly more specialists.

Since 2005, with the clear prospect of Romania's accession to the European Union, the conference takes on an international character and changes its name to ICAMaT - International Conference on Advanced Manufacturing Technologies. Ever since this Conference session there has been a continuous increase in the number of participants from home and abroad, as well as in the number of published papers.

The Conference successfully brought together engineers, researchers and professors, specialists in various fields, such as Product Development; CAD, CAM, CAE, CIM Techniques and Concurrent Engineering; Cutting Processes, Cutting Tools and Forming Technologies; Manufacturing Technologies; Technological Equipment and Automation; Management, Marketing, Production Planning and Production Quality.

31 papers were initially entered in the conference and, after reviews, 21 of them were accepted and presented. The Editors are thankful to the authors for their valuable contributions and to the reviewers for their timely reviews and recommendations as improvement guidance for selected papers. We thanks also the members of the Organizing Committee, for the consistent and substantial effort of organizing this conference.

Prof. Tom SAVU, lect. Manuela - Roxana DIJMĂRESCU, lect. Ioan - Cristian TARBĂ, Editors

IOP Conf. Series: Materials Science and Engineering **682** (2019) 011002 doi:10.1088/1757-899X/682/1/011002

Peer review statement

All papers published in this volume of *IOP Conference Series: Materials Science and Engineering* have been peer reviewed through processes administered by the proceedings Editors. Reviews were conducted by expert referees to the professional and scientific standards expected of a proceedings journal published by IOP Publishing.

IOP Conf. Series: Materials Science and Engineering **682** (2019) 012001 doi:10.1088/1757-899X/682/1/012001

On the estimation of the reliability probabilistic model of railway wheelset

M A Spiroiu[1] and M Nicolescu[2]

[1]University POLITEHNICA of Bucharest, Department of Railway Rolling Stock, Splaiul Independenţei 313, Bucharest, Romania
[2]Romanian Railway Investigating Agency, Calea Griviţei 393, Bucharest, Romania

marius.spiroiu@upb.ro

Abstract. In the present paper is made a comparison between various estimation methods for the reliability probabilistic model of the railway wheelset. The analysis is based on real field data, observed in actual operations of freight wagons. The Weibull distribution is used to model the reliability and the applied estimation methods are the Maximum Likelihood Method and the Regression Method.

1. Introduction

A railway wheelset consists of a rigid assembly of two wheels on the axle. Besides the traditional functions performed by axles and wheels on all land transport systems - supporting the vehicle weight, ensuring vehicle rolling, transmitting its traction and braking forces, etc. – the railway wheelset performs also the function of guiding the vehicle on the track, thus has an essential role in terms of traffic safety [1,2]. As shown in [3], the failure of axles and wheels are one of the main categories of rolling stock failures that have produced derailments in recent years both on the European and Romanian railway networks. Taking into account that train derailment may have important social and/or economic implications and - in the worst-case scenario – may lead to important material damage or even to fatalities [4], the wheelset is the most critical subsystem of railway vehicles in terms of the reliability [2].

Being such an important issue - both from technical and economic point of view - the railway wheelset reliability is the subject of many researches. For example, the fatigue crack issue is a subject often addressed in scientific papers, such as [5], where is presented a study on the methodology of fatigue reliability degradation of railroad wheels, considering both fatigue crack initiation and crack propagation life. Also, in [6] is investigated the influence of wheel material structure and properties on the wheel rim cracking caused by the fatigue process. The wheel-rail contact phenomena are another source of wear and damages for wheelsets, an investigation on this subject being carried out in [7].

An analysis of freight car wheelset failure modes based on field data is made in [1] and a reliability model is deduced using the regression method. Also, in [2] are analysed the railway wheel failure modes and, for the major four failure modes, is carried out an additional analysis regarding the evolution in time of the failure rates, as well as the estimation of the failure probabilistic models, based on the Weibull distribution.

IOP Conf. Series: Materials Science and Engineering **682** (2019) 012001 doi:10.1088/1757-899X/682/1/012001

The present paper aim is to make a comparison between the results obtained by using different estimation methods for the reliability probabilistic model of the railway wheelset. Reliability mathematical model is based on the Weibull distribution and its parameters are estimated on the basis of real field data.

2. Estimation methods for the reliability model parameters

The Weibull distribution is widely used in the reliability field because of its versatility. Usually it is used the two-parameter version of Weibull distribution, according to which the evolution in time of the reliability $R(t)$ is given by:

$$R(t)=\exp\left[-\left(\frac{t}{\eta}\right)^{\beta}\right], \tag{1}$$

where t is the time, β and η are the shape and the scale parameters, respectively. The versatility mentioned above is given by the structure of the probabilistic law (i.e. by the two parameters), which makes it suitable to model the failure law of practically any industrial product.

To estimate the Weibull reliability model implies thus finding the values of the shape and the scale parameters so that the model is appropriate to the experimental data. Among the different estimation methods, the most commonly used are the Regression (Least Squares) Method and the Maximum Likelihood Method. Both methods are producing good results from the point of view of modeling the real failure laws and they are applicable both for complete data sets (i.e. when the failure time is known for all the units of the observed sample) and censored data sets (i.e. when the failure time is known only for a part of the units of the observed sample), the latter being the case of this paper.

2.1. Regression Method

The Regression Method uses a linearized form of the Weibull law and the Least Squares principle in order to estimate the parameters of the Weibull reliability model. The linear form is obtained by applying twice the natural logarithm in both members of equation (1), with the result:

$$\ln\ln\frac{1}{R(t)}=\beta\ln t-\beta\ln\eta, \tag{2}$$

so the Weibull probability law can be written under a linear form:

$$Y=a_0+a_1X \tag{3}$$

where

$$X=\ln t;\ Y=\ln\ln\frac{1}{R(t)}. \tag{4}$$

Consequently, to estimate the parameters of Weibull distribution, can be used the linear regression and the Least Squares Method, which implies to minimize the sum of the squares of the deviations:

$$S=\sum_{i=1}^{n}(y_i-a_0-a_1x_i)^2 \tag{5}$$

where

$$x_i=\ln t_i\ ;\ y_i=\ln\ln\frac{1}{1-F(i)} \tag{6}$$

n being the sample size, t_i the failure time of wheelset i and $F(\mathrm{i})$ an estimation of the failures cumulative distribution function.

Using the conditions for minimizing the expression in equation (5):

$$\frac{\partial S}{\partial a_0} = \frac{\partial S}{\partial a_1} = 0 \tag{7}$$

and solving the corresponding system, the expressions for coefficients a_1 and a_0 are obtained:

$$a_1 = \left(\sum_{i=1}^{n} x_i y_i - \bar{y}\sum_{i=1}^{n} x_i\right)\left(\sum_{i=1}^{n} x_i^2 - \bar{x}\sum_{i=1}^{n} x_i\right)^{-1} ;\ a_0 = \left(\bar{y}\sum_{i=1}^{n} x_i^2 - \bar{x}\sum_{i=1}^{n} x_i y_i\right)\left(\sum_{i=1}^{n} x_i^2 - \bar{x}\sum_{i=1}^{n} x_i\right)^{-1} \tag{8}$$

where

$$\bar{x} = \frac{1}{n}\sum_{i=1}^{n} x_i\ ;\ \bar{y} = \frac{1}{n}\sum_{i=1}^{n} y_i\ . \tag{9}$$

Therefore, the estimations of parameters β and η are:

$$\hat{\beta} = a_1\ ;\ \hat{\eta} = \exp\left(-\frac{a_0}{a_1}\right). \tag{10}$$

The relationships above apply to the complete data sets (i.e. all n failure times are known). In the case of a censored data set (i.e. only k of the n wheelsets have failed, so only k failure times are known) the sums in equations (5) and (8) have the upper limit k instead of n and the means in equations (9) are calculated only for the corresponding k terms.

A very important issue within the regression method is related to the estimation of the failures cumulative distribution function in equation (6). In most cases this estimation is made using either the mean ranks or the median ranks. In the case of Mean Ranks Regression, the cumulative distribution function is given by [4]:

$$F(i) = \frac{i}{n+1} \tag{11}$$

or

$$F(i) = \frac{i-0.5}{n}, \tag{12}$$

while in the case of Median Ranks Regression is used the Benard's approximation, according to which the cumulative distribution function is estimated by the following expression [8]

$$F(i) = \frac{i-0.3}{n+0.4}\ . \tag{13}$$

2.2. Maximum Likelihood Method.

The principle of the method consists in determining the most likely estimators (i.e. that maximize the likelihood of occurrence of the empirical data set) for the unknown parameters of a distribution. Considering a complete set of experimental data $x_1, x_2, \dots x_n$, the likelihood function is defined as [4,9]:

$$L(x_1, x_2, \dots, x_n; \theta) = f(x_1;\theta)\cdot f(x_2;\theta)\cdot\dots\cdot f(x_n;\theta) = \prod_{i=1}^{n} f(x_i;\theta), \tag{14}$$

where θ is the parameter to be estimated (there may be more than one) and f is the probability density function of the assumed distribution.

The function above expresses the probability of observing the data set $x_1, x_2, \ldots x_n$ ("how likely" it is to obtain these data values) for a given θ. The estimator that is most likely to have generated the observed sample is the one that maximizes the likelihood function, so it can be found by solving the equation [4,9]:

$$\frac{dL(x_1, x_2, \ldots, x_n; \theta)}{d\theta} = 0 \ . \tag{15}$$

For practical reasons, usually is used the natural logarithm of the likelihood function instead of the function itself, so the equation to be solved is [4,9]:

$$\frac{d \ln L(x_1, x_2, \ldots, x_n; \theta)}{d\theta} = 0 \ . \tag{16}$$

In the situation of a censored data set, when for the n observed units only the first k failure times are known ($k<n$), the likelihood function is defined as [9]:

$$L(x_1, x_2, \ldots, x_n; \theta) = \prod_{i=1}^{k} f(x_i; \theta) \prod_{j=1}^{n-k} R(x_j; \theta), \tag{17}$$

where R is the reliability function.

Considering the case of Weibull distribution, the probability density function is given by:

$$f(t) = \frac{\beta}{\eta}\left(\frac{t}{\eta}\right)^{\beta-1} \exp\left[-\left(\frac{t}{\eta}\right)^{\beta}\right], \tag{18}$$

thus the likelihood function for a censored data set of k failure times given by equation (17) becomes:

$$L(t_1, t_2, \ldots, t_k; \beta, \eta) = \prod_{i=1}^{k} \frac{\beta}{\eta}\left(\frac{t_i}{\eta}\right)^{\beta-1} \exp\left[-\left(\frac{t_i}{\eta}\right)^{\beta}\right] \prod_{j=1}^{n-k} \exp\left[-\left(\frac{t_j}{\eta}\right)^{\beta}\right]. \tag{19}$$

Applying the natural logarithm:

$$\ln L(t_1, t_2, \ldots, t_k; \beta, \eta) = k \ln \beta - k\beta \ln \eta + (\beta - 1)\sum_{i=1}^{k} \ln t_i - \sum_{i=1}^{k}\left(\frac{t_i}{\eta}\right)^{\beta} - (n-k)\left(\frac{t_k}{\eta}\right)^{\beta}, \tag{20}$$

where t_k – the last observed failure time – is the censoring time, thus the time assigned to the *(n-k)* units which have survived the time t_k.

The maximum conditions are

$$\frac{\partial \ln L(t_1, t_2, \ldots, t_k; \beta, \eta)}{\partial \beta} = 0 \quad \frac{\partial \ln L(t_1, t_2, \ldots, t_k; \beta, \eta)}{\partial \eta} = 0 . \tag{21}$$

The condition above are leading, after some calculations, to the following expressions of the estimators for β and η:

$$\hat{\eta} = \left(\frac{\sum_{i=1}^{k} t_i^{\hat{\beta}} + (n-k) t_k^{\hat{\beta}}}{k}\right)^{1/\hat{\beta}} \tag{22}$$

$$\frac{1}{\hat{\beta}}+\frac{1}{k}\sum_{i=1}^{k}\ln t_i-\frac{\sum_{i=1}^{k} t_i^{\hat{\beta}}\ln t_i+(n-k)t_k^{\hat{\beta}}\ln t_k}{\sum_{i=1}^{k} t_i^{\hat{\beta}}+(n-k)t_k^{\hat{\beta}}}=0. \tag{23}$$

3. Results and discussions

In this section is performed a comparative analysis of the results obtained regarding the estimation of the reliability probabilistic model of railway wheelset, using the previously presented methods. The analysis is based on real field empirical data, observed in actual operations of freight wagons in the period of their useful (normal) life. The sample size was n=1802 units, of which there have been observed failures in the case of k = 190 units over the observation period of one year.

The Weibull parameters of the reliability model are estimated using the Maximum Likelihood Method (ML) and three variants of the Regression Method, the difference between them consisting in how is estimated the failures cumulative distribution function: by Mean Ranks Regression - given by equations (11) (MnR_1) or (12) (MnR_2) or by Median Ranks Regression (MdR).

In the case of the Regression Method it is sufficient to simply apply the procedure described above, but in the case of the Maximum Likelihood Method the two estimates are determined using iterative procedures, by considering an initial value for the form parameter in equation (23) and to adjust this value until the expression becomes null. As initial value is used one of the values previously determined through Regression Method. Once the form parameter estimation found, it is inserted in equation (22) to find the estimation of the scale parameter.

In table 1 are presented the estimations of the parameters of the Weibull reliability model for the four considered cases.

Table 1. Estimated parameters of Weibull distribution.

	$\hat{\beta}$	$\hat{\eta}$ (days)
MnR_1	1.03073	2633.93
MnR_2	1.06424	2425.38
MdR	1.04940	2514.61
ML	0.94632	3535.66

The Weibull shape parameter β, has a special significance, its value setting the monotony of the failure rate. For β =1 the failure rate is constant, while for values higher and lower than 1, the failure rate is increasing and decreasing, respectively. It can be seen in Table 1 that all β values are close to 1, so the wheelset failure rate has not important variations over time (i.e. is almost constant), these results confirming the general theory regarding the behaviour of systems during their normal life.

It can be seen that there are no major differences between the results obtained for both parameters by using the three versions of the Regression Method. Regarding the results obtained through the Maximum Likelihood Method, there is a notable difference concerning the scale parameter, which is significantly higher. On the other hand, the scale parameter is about 10% lower than those obtained by regression. It is worth noting that these results indicate that the use of the Maximum Likelihood Method leads to a more optimistic model of the wheelset reliability.

The question that arises is which of the estimated models is most appropriate for the experimental data. To answer that, in this paper is used the root mean square error (RMSE) to overall assess the deviation of the estimated reliability models in relation to actual values:

$$RMSE = \sqrt{\frac{\sum_{i=1}^{k}\left[R(t)-\hat{R}(t)\right]^2}{k}}. \tag{24}$$

In table 2 are presented the RMSE of the four estimated models of wheelset reliability. It can be seen that there are not very important differences between the four models in terms of model accuracy. It is, however, to be noted that the model obtained by the Maximum Likelihood Method proves the best fit to the empirical data, thus it seems to be the most appropriate.

Table 2. RMSE of estimated reliability models.

	RMSE
MnR_1	$2.806 \cdot 10^{-2}$
MnR_2	$2.885 \cdot 10^{-2}$
MdR	$2.849 \cdot 10^{-2}$
ML	$2.282 \cdot 10^{-2}$

4. Conclusion

The aim of the present paper was to estimate the reliability probabilistic model of railway wheelset, acknowledged as the most critical subsystem of railway vehicles from reliability point of view, and to make a comparison between various estimation methods or variants thereof. The analysis was based on real field empirical data, observed for one year in actual operations of freight wagons, in the period of their useful life.

The estimation of the Weibull reliability models was made using the Maximum Likelihood Method and three variants – given by the manner of estimation of failures cumulative distribution function - of the Regression Method. The shape parameter estimations were close to 1 in all four cases, confirming an almost constant failure rate of railway wheelset – a usual (expected) behaviour during the normal life. The scale parameter estimation through the Maximum Likelihood Method was significantly higher than those obtained through the Regression Method.

The results of the estimation indicated that the Maximum Likelihood Method leads to the most optimistic model of the wheelset reliability, however the RMSE analysis indicated this model as the most suitable for the field data. The models obtained by using Regression Method proved also good accuracy so, although the Maximum Likelihood Method seems to be the most suitable for the analysed case, it is difficult to claim that it is the appropriate method in all situations.

5. References

[1] Spiroiu MA 2015 *Appl. Mechan. Mater.* **809-810** 1097-1102
[2] Spiroiu MA and Nicolescu M 2018 *MATEC Web of Conferences* **178** 06005
[3] Nicolescu M and Spiroiu MA 2019 *U.P.B. Sci. Bull. Series D* **81**(3) 61-68
[4] Spiroiu M 2006 *Reliability and maintenance of railway vehicles (in Romanian)* (Bucharest: MatrixRom)
[5] Liu Y, Liu L, Stratman B and Mahadevan S 2008 *Reliab. Eng. Syst. Safe.* **93** 456-67
[6] Liu ZX and Gu HC 2000 *J. Mater. Eng. Perform.* **9** 580-84
[7] Mazzu A, Solazzi L, Lancini M, Petrogalli C, Ghidini A and Faccoli M 2015 *Wear* **342-343** 22-32
[8] McCool JI 2012 *Using the Weibull Distribution: Reliability, Modeling and Inference* (Hoboken: John Wiley & Sons)
[9] Rausand M and Hsyland A 2004 *System Reliability Theory Models, Statistical Methods and Applications* (Hoboken: John Wiley & Sons)

IOP Conf. Series: Materials Science and Engineering **682** (2019) 012002 doi:10.1088/1757-899X/682/1/012002

An analysis of vehicle - railway exchanger interaction

M C Tudorache[1] and I Neagoe[2]

[1] Department of Railway Vehicles, University Politehnica of Bucharest, 313 Splaiul Independenţei, Bucharest, Romania
[2] Doctoral School of Transports, University Politehnica of Bucharest, 313 Splaiul Independenţei, Bucharest, Romania

cristi_tudorache2003@yahoo.com

Abstract. The fast variation of the section and contact characteristics of railway exchangers is a difficult matter for an analytical approach. At the same time, vehicle - railway exchanger system includes multiple contact patches, discontinuities in the contact points evolution and the necessity to compute stresses and deformations. As a consequence of the harrs working conditions in the system there is a permanent concern to improve its performance. The analysis of the state of art in the research indicated the need of a complex theory for the study of the mechanical contact. In the paper, a tridimensional analysis of the vehicle - railway exchanger contact patch is carried out using "Contact" software, based on Kalker's theory. The outcome of the study is the computation of tangential forces, creepage speeds and stresses distribution in the contact patches, in order to elucidate the vehicle evolution during this interaction and the estimation of the running safety.

1. Introduction

The contact with the track devices is analyzed in three hypostases: the contact with the tongue contact [1], the contact with the crossing heart and the contact with the rail which requires great efforts in support plates, which is why it cracks. The countermeasure plays an essential role in guiding latency over the track device.

Hiroyuki Sugiyama [2] presents a numerical analysis of the non-conformal wheel-decal contact. It parameterizes the contact surfaces and draws a parallel between the nodal analysis and the non-conforming contact, highlighting its efficiency from the perspective of an online analysis. The method is quite expensive.

According to the analyzed bibliography, many methods of analyzing the behavior of the vehicle were developed when passing over the track devices. Therefore, simulations performed using the GENSYS software have been experimentally validated by tests conducted on a convoy scale [3].

The finite element approach has been approached in more than 500 ways for analyzing the dynamic behavior of the vehicle when passing over the track devices [4]. The simulation of the contact using the SIMPACK package is described in the reference [5-6]. In reference [7-8], the authors used several Euler-Bernoulli beams in the path model to simulate the rails in the runoff zone over the track changer and in accordance with the path structure. Another model of analysis constitutes the model of real-time contact realized in NUCARS in which a bearing is used carriages on some lanes of passage

Another method of analyzing the behavior of the vehicle when passing over the track devices is the VAMPIRE package [9]. To determine the contact forces with the help of this software there are four stages: - solid modeling of the conventional system of interaction is performed - the points of contact

IOP Conf. Series: Materials Science and Engineering **682** (2019) 012002 doi:10.1088/1757-899X/682/1/012002

between the axle and the running path are established, - the trajectory of the contact point is analyzed when crossing the crossing heart - dynamic behavior is analyzed (contact forces).

It is known that the track devices represent an unfavorable side of the tread and one of the most important factors limiting the speed when passing over them. For this reason it is necessary to analyze the dynamic characteristics at the contact level in order to improve the structure, to reduce the transverse and longitudinal shocks and mainly to increase the speed of movement in this area.

Zhai and Ren [10], Zhao [11] and Andersson and Dahlberg [12] studied the behavior of vertical direction of the interaction between the wheel and the track changer, analyzing the relationship between the speed of movement and the dynamic vertical forces. Wang [13] states in 1997 that a study of the dynamic characteristics of the wheel-changer system on a simplified model cannot be carried out precisely.

Drozdzie [14] analyzed the geometry of the tread deviations but does not include the dynamic effect of the vehicle. Baluch [15] analyzes the kinematic behavior of the vehicle when crossing the discontinuities of the track but does not include the dynamic effect, [16]

The pressure in the wheel-contact area is determined by the size of the contact force and the dimensions of the contact area. According to the specialized studies it is found that the pressure in the contact area does not significantly influence the speed of movement if the shape of the contact surfaces and the normal load on the wheel are taken into account, a fact stated by E. Kassa and G. Johansson. Thus, for the nominal profile of the wheel it is found that the pressure increases with the load and for a worn profile it reaches more maximums which contributes to reducing the wear [17].

Finally, when analyzing the specialized bibliography it is observed that most of the articles analyze the case of the vehicle's interaction with the tip of the crossing heart, so it is necessary the need for a three-dimensional analysis of the wheel contact with the tongue of the wheel in terms of traffic safety. This analysis is performed with the iterative software Contact - Kalker adapted for bicontact and the results are graphically visualized through the Matlab programming system.

2. The problem of wheel contact – tread

Since 1967 Kalker, and later other authors, have developed different theories specifically dedicated to wheel-rail contact analysis. These theories consider that the two bodies have identical elastic characteristics and respect Hertz's theory regarding the normal distribution of force. This allows the study of the contact to be divided into two independent problems: the problem related to the description of the normal force, the so-called normal problem and the problem related to the description of the tangential forces, the so-called tangential problem. According to Kalker's theory, the dependence of the tangential forces on the pseudo-slip in the contact area is given by the linear expressions:

$$F_x = -a \cdot b \cdot G \cdot C_{11} \cdot \nu_x,$$

$$F_y = -a \cdot b \cdot G\left(C_{22} \cdot \nu_y + \sqrt{a \cdot b} \cdot C_{23} \cdot \phi\right) \tag{1}$$

$$M_\phi = -\left(\sqrt[3]{a \cdot b} \cdot G \cdot C_{32} + (a \cdot b)^2 G \cdot C_{33} \cdot \phi\right)$$

where: a, b - the semiaxes of the contact ellipse, ν_x, ν_y, φ - the longitudinal, transverse and spin pseudo-slip at the contact points, C_{11} , C_{22}, $C_{23,32}$, C_{33} - Kalker's coefficients, which depend on the value of the coefficient Poisson being established based on the relationships below [18]:

$$C_{11} = -\frac{\partial F_x}{\partial \nu_x} / (a \cdot b \cdot G) \quad C_{22} = -\frac{\partial F_y}{\partial \nu_y} / (a \cdot b \cdot G) \quad C_{23} = -\frac{\partial F_y}{\partial \phi} / \left(\sqrt[3]{a \cdot b} \cdot G\right)_{\nu_x = \nu_y = \phi = 0} \tag{2}$$

IOP Conf. Series: Materials Science and Engineering **682** (2019) 012002 doi:10.1088/1757-899X/682/1/012002

The values of the semiaxes of the contact ellipse are calculated with the relation, [19]:

$$a = m\left(\frac{3}{2}N\frac{1-\upsilon^2}{E(A+B)}\right)^{1/3}$$
$$b = n\left(\frac{3}{2}N\frac{1-\upsilon^2}{E(A+B)}\right)^{1/3} \qquad (3)$$

where: m, n are sizes that depend $cos\,\beta = \frac{|B-A|}{B+A}$ on Hertz, $N = \iint_S p_n(x,y)\,dxdy$ - normal wheel load.

Kalker also defines the pseudo-slip coefficients as follows:

$$\kappa_x = \frac{a\cdot b\cdot G\cdot C_{11}\cdot \nu_x}{\mu\cdot N}$$
$$\kappa_y = \frac{a\cdot b\cdot G\cdot C_{22}\cdot \nu_y}{\mu\cdot N} \qquad (4)$$
$$\kappa_\phi = \frac{\sqrt[3]{a\cdot b}\cdot G\cdot C23\cdot \phi}{\mu\cdot N}$$

According to [20], the modulus of transverse elasticity takes into account the nature of the material from which the rail and the wheel are made, so that it is given by the relation:

$$G = \frac{2\cdot G_r\cdot G_s}{G_r + G_s}$$

And Poisson's coefficient is dependent on the elastic characteristics of the material

$$\upsilon = \frac{G\cdot(G_r\cdot\upsilon_s + G_s\cdot\upsilon_r)}{2\cdot G_r\cdot G_s}$$

The longitudinal, transverse and spin pseudowords are established based on the relationships, [19]:

$$\nu_x = \frac{w_x}{V},\quad \nu_y = \frac{w_y}{V},\quad \phi = \frac{r\cdot\omega_s}{V}, \qquad (5)$$

where ω_s – angular velocity of spin.

IOP Conf. Series: Materials Science and Engineering **682** (2019) 012002 doi:10.1088/1757-899X/682/1/012002

3. Numerical simulations

Within this section we analyze the behavior of an axle equipped with profile wheels S1002 (nominal rolling radius 460 mm) when passing over a simple track changer made from rail profile of type UIC 60. Axle speed over the track changer it is 100 km/h. The contact is analyzed from a three-dimensional point of view. Normal wheel load is considered to be 15-20 kN.

The profiles were drawn by the circle arcs method and the contact points were established by the minimum distance method. A gap of 6 mm was considered and two contact points were identified as follows: one on the lip of the wheel and one on the tread surface.

Starting from the idea that the tangential forces are transmitted by the adhesion between the two interacting bodies, which is dependent on the elastic deformations and the slips in the contact area.

At the same time, the pseudo-slip is a parameter that influences the transmission of the tangential forces in the wheel-rolling contact area, being defined by the ratio between the sliding speed and the axle advancement speed. The following shows the vectorial distribution of the tangential tensile force relative to the normal load on the wheel.

It can be observed that this increases with its vector field as the load increases, see figure 1-2.

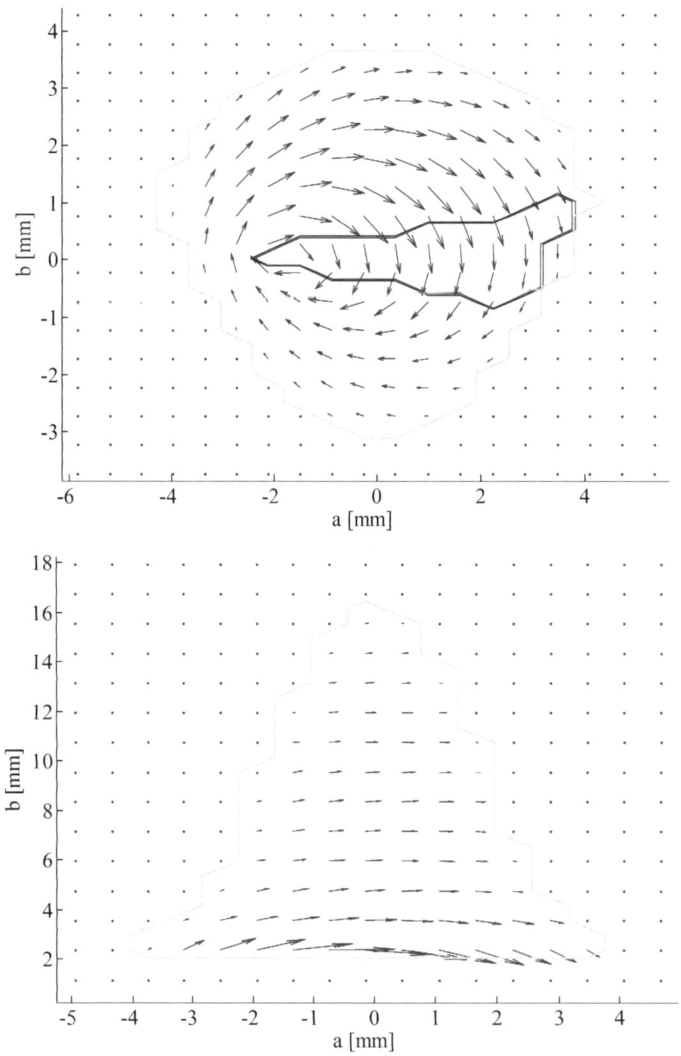

Figure 1. Vector distribution of the tangential tensile force for the normal load of 15kN

At the contact point on the lip, according to the distribution of the tangential force field of traction, the wheel runs on the rail similar to the running situation without pivoting slips. In this case, the intensity of the traction field increases with the load, see figure 2. There are no major changes in the contact point on the tread surface.

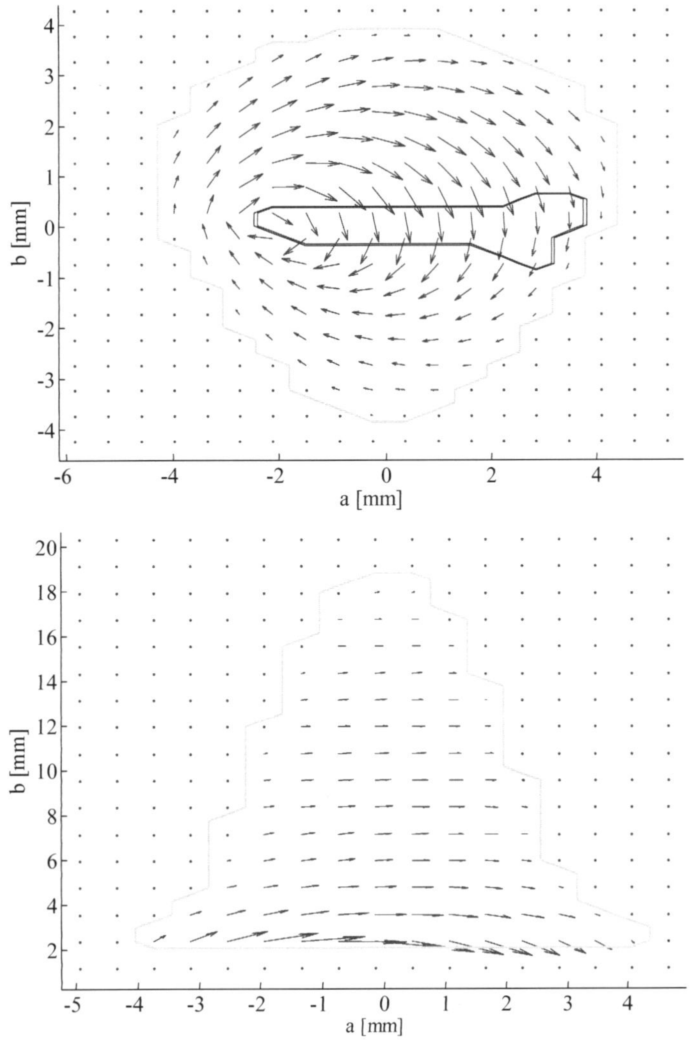

Figure 2. Vector distribution of the tangential tensile force for the normal load of 20 kN

The distribution of the slip and adhesion zones is/are analyzed in the case of the bicontact that appeared at the interaction between the wheel and the hub tongue, see figures 3-4.

It is observed that at the point on the lip the distribution of the slip and adhesion zones is similar to the rolling in the conditions where the spin pseudo-slip is zero, is the wheel behaves like a puppet, see figures 3-4.

It is observed that the size of the normal load on the wheel significantly changes both the shape and the size of the grip area in the transverse direction, which is justified by an increase in the tangential force.

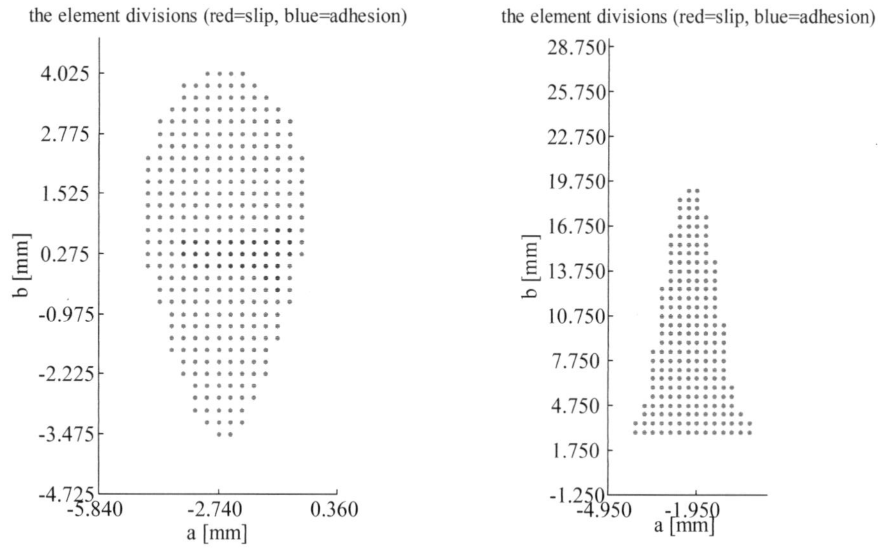

Figure 3. Distribution of the adhesion and slip zones for the normal load of 20 kN

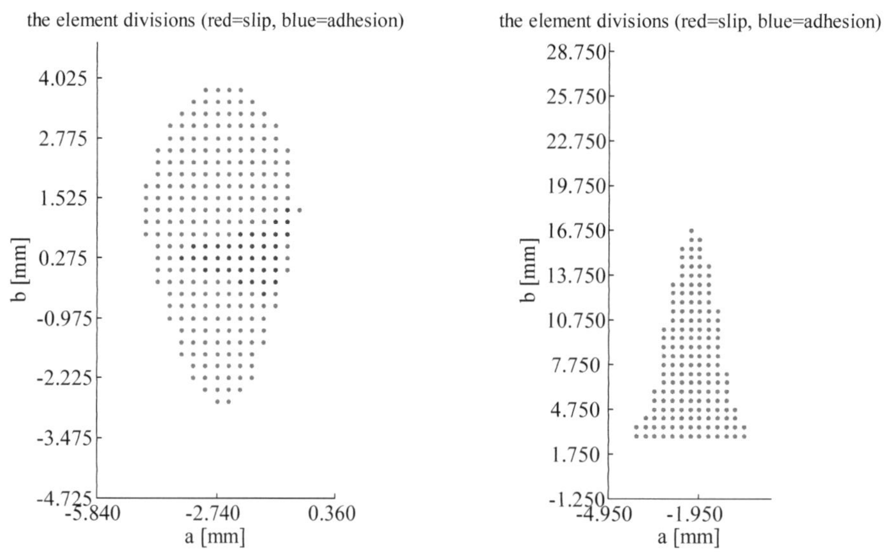

Figure 4. Distribution of grip and slip zones for the normal load of 15 kN

Figures 5 - 6 show the distribution of the normal pressure in the wheel-axle contact area according to the load on the wheel. It is found that the pressure is higher when contacting the tongue tip, which is justified by the increased wear in this area.

The normal pressure in the contact area is directly influenced by the load on the wheel. It can be seen figures 5 - 7 that the wheel load significantly influences the maximum pressure at the wheel lip contact.

IOP Conf. Series: Materials Science and Engineering **682** (2019) 012002 doi:10.1088/1757-899X/682/1/012002

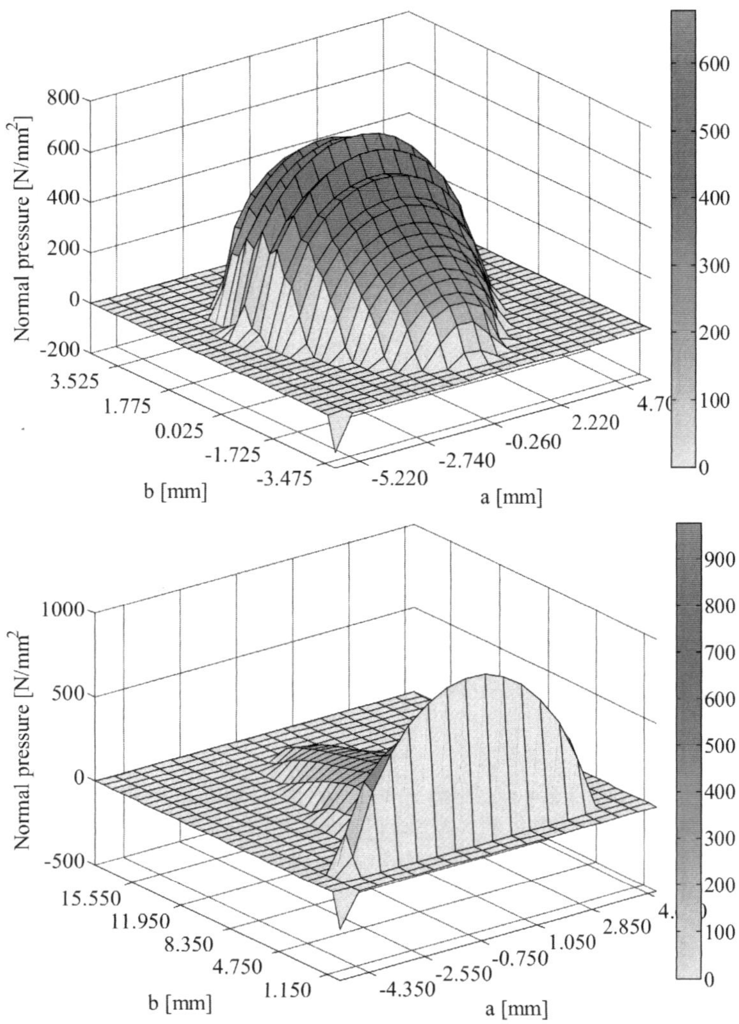

Figure 5. Normal pressure distribution for the normal load of 15 kN

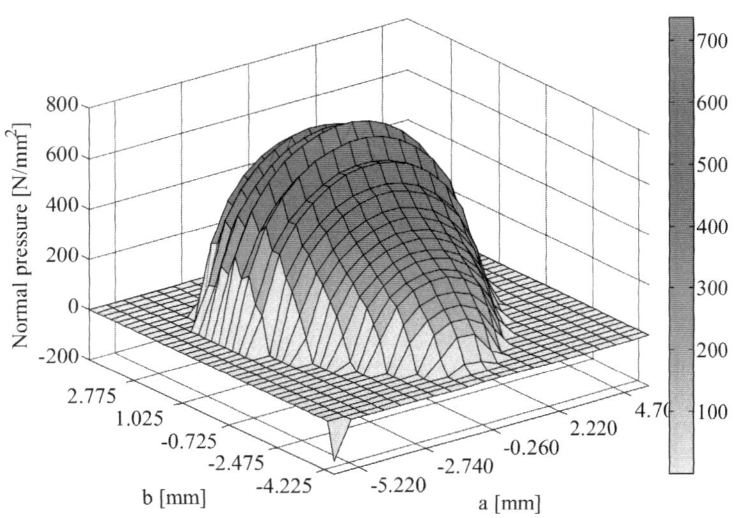

Figure 6. First point of contact - Normal pressure distribution for the normal load of 20 kN

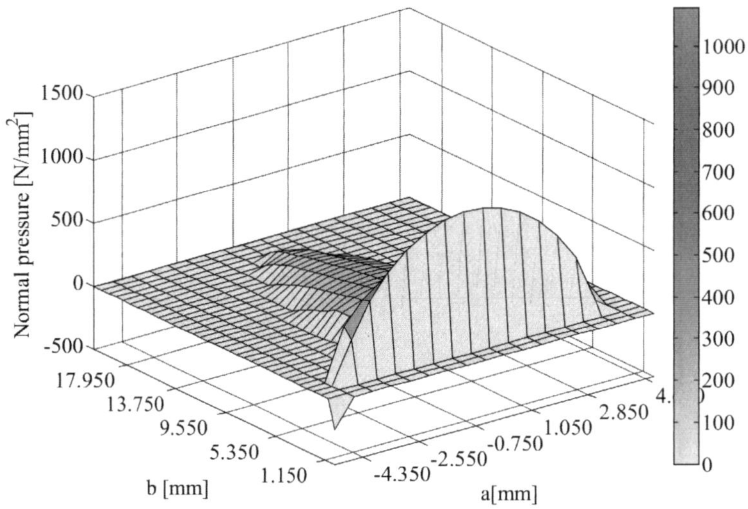

Figure 7. Second contact point - Normal pressure distribution for the normal load of 20 kN

The justification of the necessity of the study carried out is taken into account that, for the study of the wheel-rail contact phenomena, essentially for the determination of the tangential forces, the knowledge of the sliding speeds and the distribution of the pressures in the contact area is defining for the behavior of the vehicle in the lane.

4. Conclusion

A method for the wheel-turnout interaction study is developed in the present work. The tridimensional Kalker theory is employed using the computer program CONTACT. The application is designed for frictional concentrated contact problems. The solution proposed is numerically efficient and it has been previously validated for the general case of the wheel-rail contact.

Multi-point contact is a common situation for the studied system. In the approach proposed, the bodies deformations and the contact areas can be determined from the contact loading. Further, the number and position of the contact points can be calculated either on-line or off-line. The scheme proposed also allows the computation of the loading and stress state starting from the local deformations.

5. References

[1] Tudorache C, Oprea R and Stafie M 2018 *IOP Conference Series-Materials Science and Engineering* **400** 022057
[2] Sugiyama H and Sekiguchi T 2012 *Multibody Syst. Dyn.* **27** 55 –74
[3] Kassa E and Nielsen J C O 2008 *Veh. Syst. Dyn.* **46(5)** 521–534
[4] Kassa E, Anderson C and Nielsen J 2006 *Veh. Syst. Dyn.* **44** 247–258
[5] Kono H, Suda Y, Yamaguchi M, Yamashita H, Yanobu Y and Tsuda K 2005 *Proc. of ASME Int. Design Engineering Technical Conf. and Computer and Information in Engineering Conf.* (USA)
[6] Alfi S and Bruni S 2008 *Veh. Syst. Dyn.* **47** 551–574
[7] Ren Z, Sun S and Zhai W 2005 *Veh. Syst. Dyn.* **43(4)** 285–303
[8] Shu X, Wilson N, Sasaoka C and Elkins J. 2006 *Veh. Syst. Dyn.* **44(Suppl)** 251–260
[9] Sun Y Q, Cole C and Boyd P 2010 *Wear* **271(1-2)** 482–491
[10] Zhai W M and Ren Z S 1998 *J. of the China Railway Society* **20** 30–38
[11] Zhao X 1998 *The fixed switch analysis of FEA and the elastic parameter optimizing*, MSc thesis (China: China Railway Research Institute)
[12] Andersson C and Dahlberg T 2000 *Veh. Syst. Dyn.* **33** 131–142

[13] Wang P 1997 *Research on wheel/rail system dynamics on tournout*, PhD thesis, Southwest Jiaontong University of China
[14] Drozdzie J, Sowinski B and Groll W 1997 *Proc. of 16th IAVSD Symposium* (Pretoria, South Africa)
[15] Baluch A M 1998 *Archives of Civil Engineering* **44** 301–315
[16] Zhai W M and Sun X 1994 *Veh. Syst. Dyn.* **23** 603–615
[17] Kassa E and Johansson G 2006 *Veh. Syst. Dyn.* **44(sup1)** 349–359
[18] Kalker J J 1991 *Wear* **144** 243–261
[19] Sebeșan I 1996 *Dinamica vehiculelor de cale ferată* (Bucharest: Editura Tehnică)
[20] Zougar A 2014 *Estudio del comportamiento vibratorio de vias ferroviarias mediante simulation numerica*, PhD thsesis (Barcelona: Universitat Politecnica de Catalunyia)

Influence of spin creepage on contact patch

M C Tudorache[1], R A Oprea[2] and I Neagoe[3]

[1,2] Department of Railway Vehicles, University Politehnica of Bucharest, 313 Splaiul Independenței, Bucharest, Romania
[3] Doctoral School of Transports, University Politehnica of Bucharest, 313 Splaiul Independenței, Bucharest, Romania

cristi_tudorache2003@yahoo.com, razvan.oprea@gmail.com

Abstract. The wheel-rail contact patch analysis methods are related to the evolution of modern computers. Contemporary computation systems allow the approach of tri-dimensional methods which take into account various contact parameters. With this background in mind, the wheel-rail contact problem is a study under development. The paper proposes a tri-dimensional approach, considering the spin creepage parameters. The spin effect is due to the increase of the wear around the wheel flange and the gage face. At the same time, the spin is influenced by the wheel rolling operating conditions. The paper gives an insight of wear mechanisms analysis, plastic deformation and fatigue phenomenon of the contact patch constituent material.

1. Introduction

An important application area for the Contact Kalker software is the detailed study of the wheel-rail interaction. The software features advanced characteristics and in each element of the contact area the dry friction Coulomb law may be applied. Contact Kalker is using an iterative algorithm which gives accurate results in the railway transportation domain [1-4]. By means of the Contact algorithm one can obtain elastic deformations, stick and slip areas, creepage values, contact forces and pressures as described in figure 1.

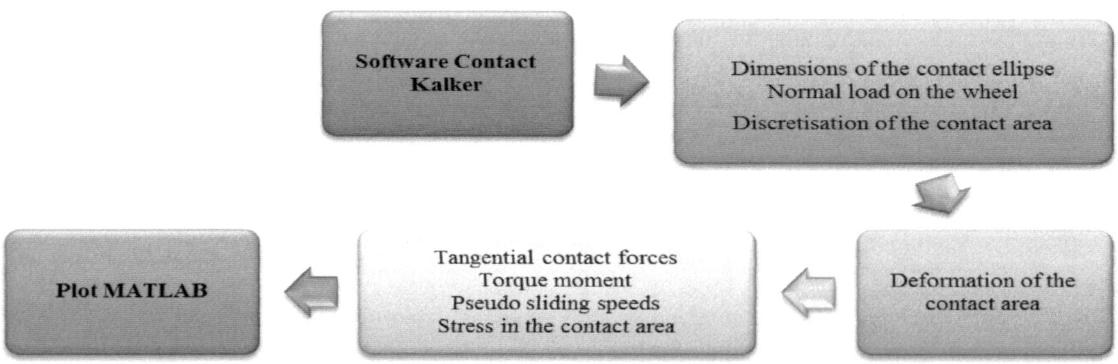

Figure 1. Contact interface [3].

The preparing stages for the input data of the Contact Kalker software are the following:
1 - obtaining the coordinates of the first contact point using the method of the minimal level difference
2 - deformation calculus considering an isolated contact point
3 - testing the possibility of a new contact area
4 - computing deformations in the contact zones
5 - examination of the vertical components values; if the differences are not within assumed tolerances, step 4 is re-iterated using vertical differences computed at step 5, until the imposed accuracy is obtained.

In the following, a mounted axle with wear profiles rolling (nominal radius 460 mm) on rails with UIC 60 profiles is analyzed using Matlab and Contact Kalker software.

2. Aspects of wheel-rail contact

At the end of the XIXth century, Heinrich Hertz [5] accomplished sustained research regarding mechanic contact between two bodies and published his first paper in 1896.

His studies were based on optic interference produced by cylindrical lenses. He mentioned that both lenses were deformed under contact stress and that the contact area is elliptical. When the stress was removed, the contact patch disappeared and the lenses regained their initial shape. Starting from this observations, Hertz stated the theory presently used on a large scale to develop contact models due to its simplicity and the satisfactory accuracy. In spite of its advantages, the theory has certain limitations:
- it may be applied only to smooth surfaces with constant curvature in the contact area
- it allows to describe only normal forces on the contact surface

Since 1967, Kalker and subsequently other authors developed different theories suited especially to the wheel-rail contact study, [6]. These theories consider that the bodies in contact have identical elastic characteristics and that they observe Kalker theory regarding the normal distribution of the force. The main contribution of Kalker's theory is that he considered the influence of the spin movement.

Kalker defines the creep coefficients as follows, [7,8]:

$$\kappa_x = \frac{a \cdot b \cdot G \cdot C_{11} \cdot \nu_x}{\mu \cdot N}, \quad \kappa_y = \frac{a \cdot b \cdot G \cdot C_{22} \cdot \nu_y}{\mu \cdot N}, \quad \kappa_\phi = \frac{\sqrt[3]{a \cdot b} \cdot G \cdot C_{23} \cdot \phi}{\mu \cdot N} \tag{1}$$

where: a, b – contact ellipsis dimensions, G – the modulus of transverse elasticity, $C_{11,22,23}$ – Kalker's coefficients, $\nu_{x,y}$, ϕ - longitudinal transverse and spin creepage.

Longitudinal transverse and spin creepage are determined from equation (2):

$$\nu_x = \frac{w_x}{V}, \quad \nu_y = \frac{w_y}{V}, \quad \phi = \frac{r \cdot \omega_s}{V}, \tag{2}$$

where: $w_{x,y}$ – longitudinal and transverse relative velocities, r – wheel rolling radius, V – wheel rolling velocity,, ω_s – spin angular velocity.

The creep velocities are given by equations. (3, 4):

$$0 = w_{x,y} = V \cdot \left(\nu_x - \phi \cdot y,\ \nu_y + \phi \cdot x\right) - V \frac{\partial u(x,y)}{\partial x} \text{ - inside the contact ellipsis} \tag{3}$$

$$0 = p(x,y) \text{ - outside contact ellipsis} \tag{4}$$

As stated in [9], the spin creepage is obtained decomposing angular velocity vector into normal and tangential components regarding the wheel-rail contact plane.

The spin effect is considerable when the contact takes place on the wheel flange. Besides that, spin creepage is significantly increased when the wheel is acted by a traction torque. This effect also increases wear of powered axles wheels.

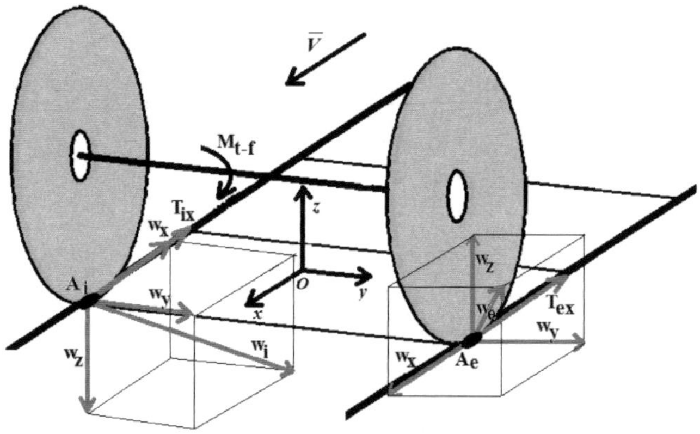

Figure 2. Longitudinal transverse and spin creepage [3].

The analyses of the axle kinematics [10] is accomplished considering a symmetric loaded axle, rolling with velocity V along a railway section either in traction or braking. The result is obtained using the torques equation (5):

$$M_{t-f} - r_e T_{ex} - r_i T_{ix} = 0 \tag{5}$$

where: M_{t-f} – traction/breaking torque, r_e, r_i – rolling radiuses in the contact point, T_{ex}, T_{ix} – tangential forces in the contact point.

The torque exerted by the axle is positive for traction and negative for breaking. It is known that M_{t-f} may not surpass $2r\mu N$ i.e., the value at the adherence limit

In this situation the longitudinal creepage velocities are given by equation (6):

$$w_{ex} = v\,(1 + e/R) - K \cdot r_e / r,$$
$$w_{ix} = v\,(1 - e/R) - K \cdot r_i / r \tag{6}$$

where v – axle rolling velocity, e – half distance between rolling radiuses, r – wheel nominal radius, r_e, r_i – rolling radiuses in the contact points, K – traction/breaking coefficient.

Tangential forces are given by the equation (7):

$$T_{ex} = -\kappa_{ex} \cdot (1 + e/R - K \cdot r_e / r) Q_e$$
$$T_{ix} = -\kappa_{ix} \cdot (1 - e/R - K r_i \cdot / r) Q_i \tag{7}$$

IOP Conf. Series: Materials Science and Engineering **682** (2019) 012003 doi:10.1088/1757-899X/682/1/012003

The torques equation (5) becomes:

$$M + r_e \cdot \kappa_{ex} \cdot (1 + e / R - K \cdot r_e / r) \cdot Q_e - .....$$

$$..... - r_i \cdot \kappa_{ix} \cdot (1 - e / R - K \cdot r_i / r) \cdot Q_i = 0 \quad (8)$$

And hence:

$$K = \frac{M + r_e \cdot \kappa_{ex} \cdot Q_e + r_i \cdot \kappa_{ix} \cdot Q_i + (e/R)(r_e \cdot \kappa_{ex} \cdot Q_e - r_i \cdot \kappa_{ix} \cdot Q_i)}{(1/r) \cdot \left(r_e^2 \cdot \kappa_{ex} \cdot Q_e + r_i^2 \cdot \kappa_{ix} \cdot Q_i\right)} \quad (9)$$

The following notations are used:

$$r_e \cdot \kappa_{ex} \cdot Q_e = r_e \cdot C_{11} \cdot a_e \cdot b_e \cdot G, \qquad r_i \cdot \kappa_{ix} \cdot Q_i = r_i \cdot C_{11} \cdot a_i \cdot b_i \cdot G$$

$$k_1 = r_e \cdot \kappa_{ex} \cdot Q_e + r_i \cdot \kappa_{ix} \cdot Q_i, \qquad k_2 = r_e \cdot \kappa_{ex} \cdot Q_e - r_i \cdot \kappa_{ix} \cdot Q_i \quad (10)$$

where: κ_{ex}, κ_{ix} - longitudinal creep coefficients, $Q_{e,i}$ – both wheels are equally loaded with load N. Considering the above notations the traction/braking coefficient becomes:

$$K = \frac{M_{t-f} + k_1 + (e/R) \cdot k_2}{(1/2 \cdot r)(r_e \cdot (k_1 + k_2) + r_i \cdot (k_1 - k_2))} \quad (11)$$

The values of the traction/braking coefficient may be as follows:

- $K = 1$ – free axle
- $0 \le K < 1$ - braking axle
- $1 \le K < \infty$ - powered axle

A null traction/braking coefficient is for a stocked wheel value tends to infinity when rolling at adherence limit.

3. Numerical simulations

In the present section the behaviour of an powered axle is analyzed. The axle wheels profile is S1002 (nominal rolling radius 460 mm) and the rail profile is UIC 60. The axle forward velocity is160 km/h. The wheel normal load is between 15-25 kN.

After completing preparing stages, bi-contact is obtained. The contact points are located on the rolling surface and the wheel flange.

The distribution of the normal load on the wheel is based on equal elasticities. The distribution of vertical components of the normal load in the contact areas is given by the vertical components of the penetrations in these areas and the resultant load is equal with the wheel load.

Taking into account the influence of the spin upon wheel flange and gage face a 3D analysis of the frictional power density is presented in the following.

As the frictional power density is proportionate to tangential stress, forward velocity and creep in the contact area [1], in figures 3, 4 the values in the two contact points are given for different wheel loads, considering that spin coefficient is 0,000923 rad/mm.

The increase of the frictional power density when wheel load increases is considerable. As the slip area is larger on the rolling surface, the values are larger compared to those on the wheel flange.

IOP Conf. Series: Materials Science and Engineering **682** (2019) 012003 doi:10.1088/1757-899X/682/1/012003

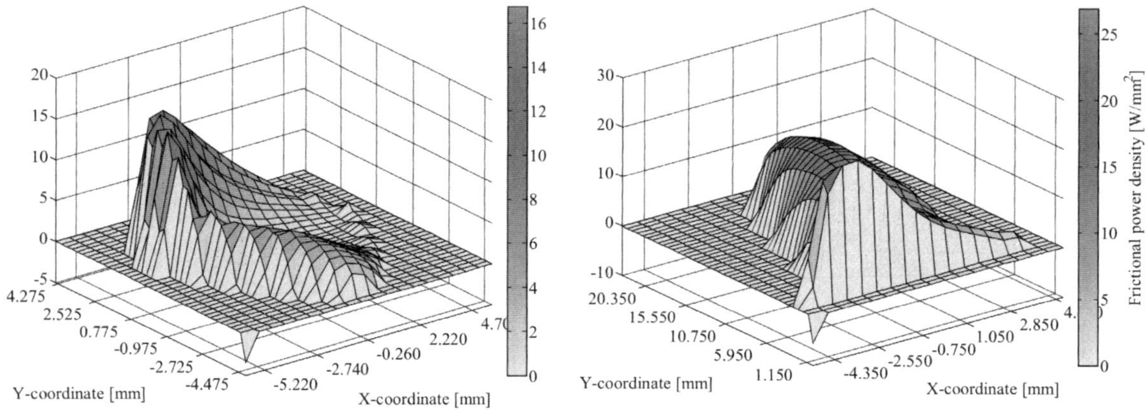

Figure 3. The distribution of the frictional power density for 25kN wheel load

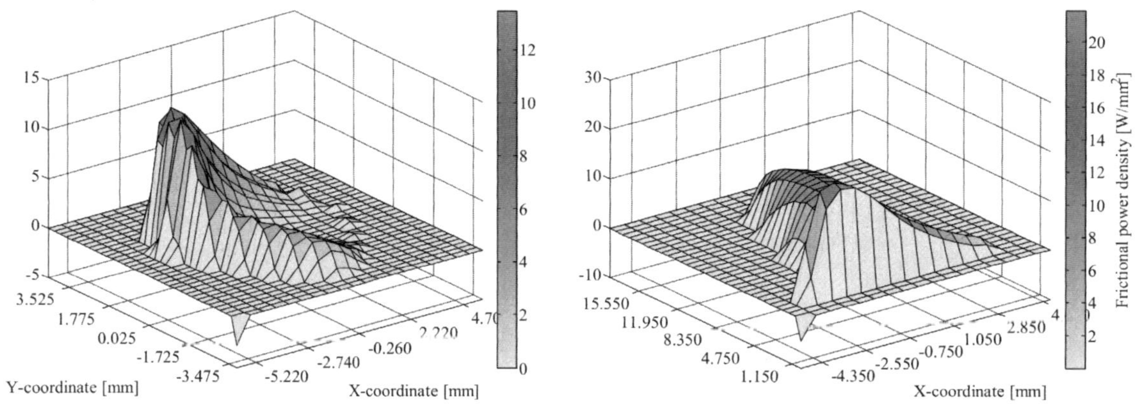

Figure 4. The distribution of the frictional power density for 15kN wheel load

It may be noticed that in figures 5-6 neglecting the spin effect, i.e., simulating a braking axle, significantly diminishes the values of the frictional power both for 25kN and 15kN wheel load, compared with the previous situations. This fact is explained by the axle tendency to slip forward.

Concurrently, on the wheel flange the shape of the stick area modifies in a similar way with the one predicted by the stripes theory. The increase in the spin rate makes the computation very difficult as in the case of the 3D theory or in the case of the simplified theory when taking spin into account makes impossible to obtain a solution. The stick area is very small compared to the slip one (bordered by the sharp edges on the plots).

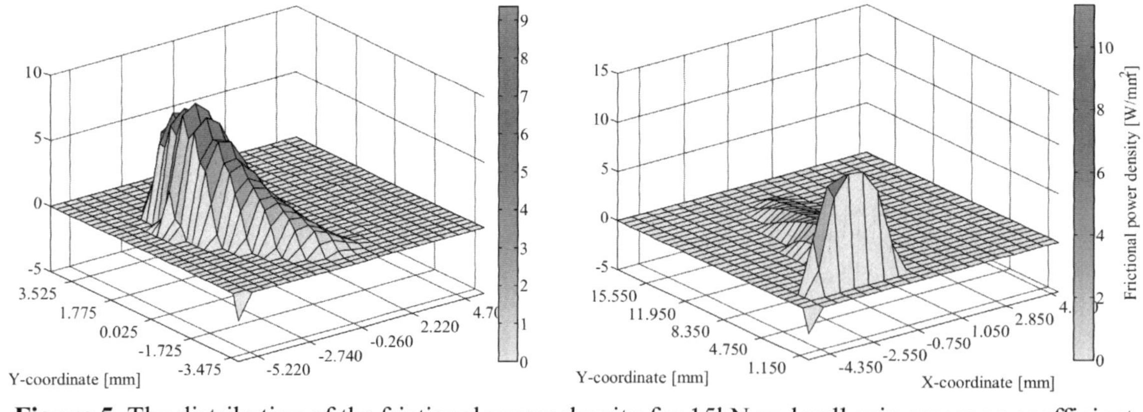

Figure 5. The distribution of the frictional power density for 15kN and null spin creepage coefficient

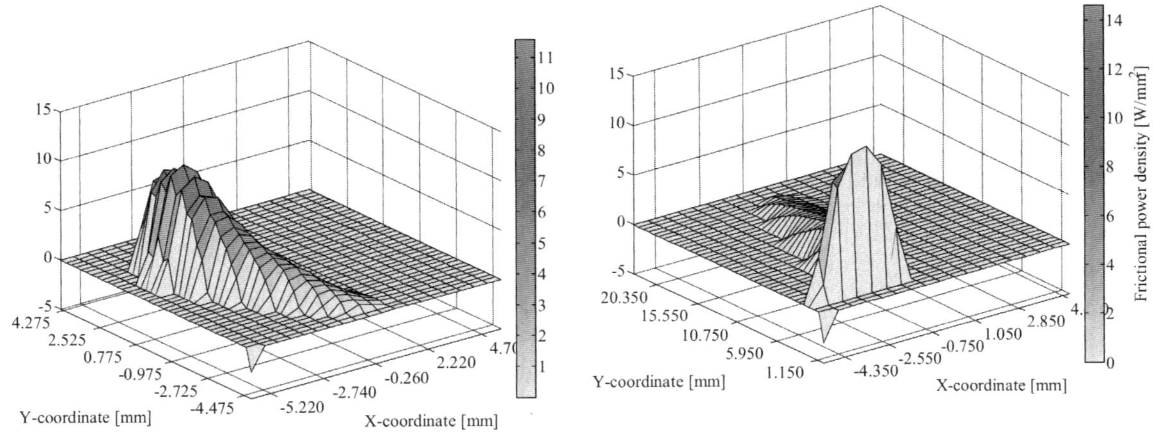

Figure 6. The distribution of the frictional power density for 25kN and null spin creepage coefficient

In figures 7-8 the frictional power density is depicted for a powered axle with 25kN wheel load. Forward running velocities are 200 km/h and 30km/h, respectively. It may be seen that increased forward speed implies and increased frictional power loss in both contact points.

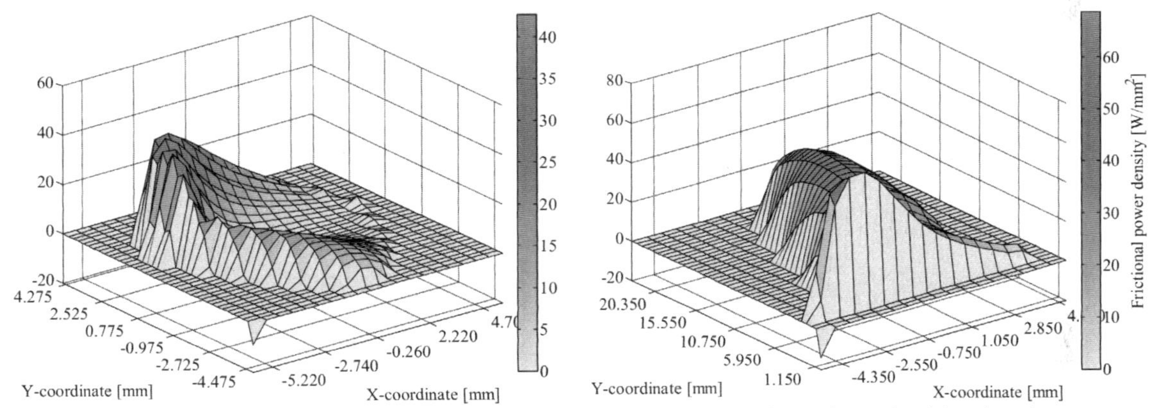

Figure 7. The distribution of the frictional power density for 25kN wheel load and forward speed 200 km/h

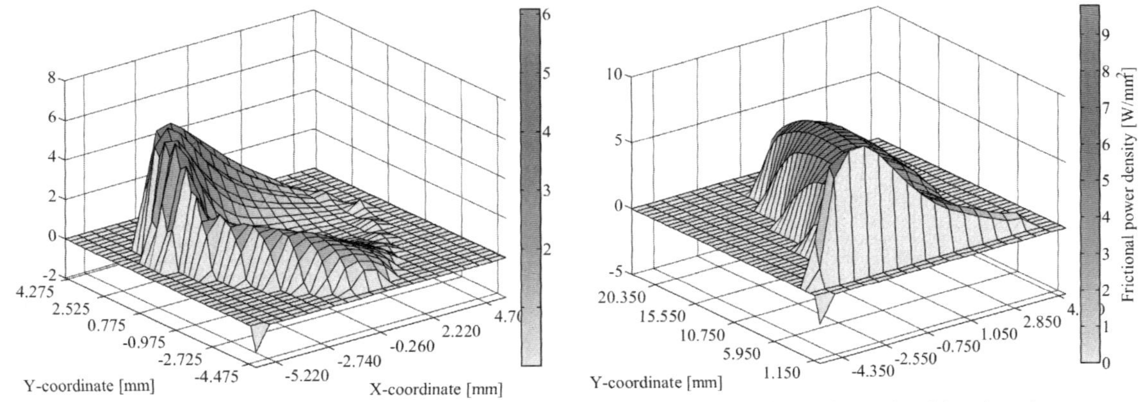

Figure 8. The distribution of the frictional power density for 25kN wheel load and forward speed 30 km/h

IOP Conf. Series: Materials Science and Engineering **682** (2019) 012003 doi:10.1088/1757-899X/682/1/012003

4. Conclusion

The paper presents a 3D analysis of wheel rail contact area considering the spin effect. The presence of the spin increases wear in wear of the wheel flange and the gage face areas. Moreover, it may be noticed that the spin effect is influenced by the axle torque when powered or when braking and further, the running safety is influenced by the axle torque.

The present study opens the perspective of the wear advance mechanism, plastic deformations and material fatigue in the wheel-rail contact area.

5. References

[1] Vollebregt E A H 2012 User guide for CONTACT, Vollebregt & Kalker's rolling and sliding contact model *VORtech Computing the Scientific Software Engineers* (The Netherlands)

[2] Oprea R A and Tudorache C 2017, *Fenomene deterministe și procese aleatoare la rularea vehiculelor feroviare* (Bucharest: Ed. MatrixRom)

[3] Sebeşan I and Tudorache C 2014 *Sci. Bull. - "Politehnica" Univ. Buchar. Ser. D: Mech. Eng.* **76(3)** 93–104

[4] Tudorache C and Oprea R A 2015 *Sci. Bull. - "Politehnica" Univ. Buchar. Ser. D: Mech. Eng.* **77** 37–50

[5] Hertz H and Lenard Ph 1895 *Gesammelte Werke: Schriften vermischten Inhalts* (Barth: Leipzing) pp 155

[6] Knothe K 2008 Veh. Syst. Dyn. **76** 9–26.

[7] Kalker J J 1991 *Wear* **144** 243–261

[8] Kalker J J 1967 *Proc. of KNAW B70* (Koninkl Nederland Akademic van Wetensharppen-Amsterdam, Reprinted from Proccedings B **70(1)** 10–62)

[9] Sebeşan I 1995 *Dinamica vehiculelor de cale ferată* (Bucharest: Ed. Tehnică) pp. 101–102

[10] Mazilu T 2008 *Vibrații roată – șină* (Bucharest: Ed. MatrixRom)

IOP Conf. Series: Materials Science and Engineering **682** (2019) 012004 doi:10.1088/1757-899X/682/1/012004

Critical points numerical analysis of ride comfort of the flexible railway carbody

D I Stănică[1] and M Dumitriu[2]

[1] Doctoral School of Transports, University Politehnica of Bucharest, 313 Splaiul Independenţei, Bucharest, Romania
[2] Department of Railway Vehicles, University Politehnica of Bucharest, 313 Splaiul Independenţei, Bucharest, Romania

madalinadumitriu@yahoo.com

Abstract. Ride comfort is one of the criteria for evaluating the dynamic behaviour in railway vehicles, through which the complex sensation triggered by the vibrations in the railway vehicle carbody upon passengers is being described. The behaviour of vibrations along the vehicle carbody is not uniform and the point where the ride comfort is the least convenient can be considered as the carbody critical point. The paper examines the ride comfort along the carbody and the position of the ride comfort critical points in correlation with the speed, carbody flexibility and suspension damping. To this purpose, there are used the results from numerical simulations regarding the ride comfort index calculated along the carbody and the ride comfort index in three points deemed relevant in terms of evaluating the ride comfort at the center of carbody and above the bogies.

1. Introduction

While train riding, the comfort of passengers can be affected by various factors, where some of them derive from the movement of the railway vehicle, such as vibrations and noise, whereas others from the environment conditions within the vehicle – temperature, humidity and air speed, lighting – or the outside fitting and facilities (e.g. the shape and placement of chairs) [1].

Of all the factors influencing the comfort of passengers, vibrations are paid a great attention since their effect upon the human body is extremely important [2 - 5]. In dependence on the intensity of vibrations, direction and frequency of occurrence, as well as the exposure time, vibrations can harm the human health or his ability to conduct certain sedentary activities. Generally, the vibrations in the railway vehicle are considered the main factor that determines in the ride comfort [6 - 8].

Ride comfort is one of the criteria for evaluating the dynamic behaviour in railway vehicles [8, 9]. This is used to describe the degree of the passengers' comfort from the point of view of mechanical vibrations, taking into account the physiological characteristics of the human body [10].

In vertical plan, the level of vibrations is not uniform along the vehicle carbody, and the point in which the ride comfort is the least convenient can be seen as the critical point of carbody when it comes to ride comfort [11]. In general, the level of vibrations is lower at the carbody center and higher at its ends. For the flexible carbodies of the high speed vehicles, the level of vibrations at the center of carbody can increase much, due to the first carbody vertical bending mode [12 - 15].

In paper is analyzed the ride comfort along the vehicle carbody and the position of the ride comfort critical points, in correlation with the speed, carbody flexibility and suspension damping. The evaluation of the ride comfort is ruled by the ride comfort index for vertical vibrations, derived from numerical simulations. The vehicle is represented by a rigid-flexible coupled model, which comprises the flexible carbody and six rigid bodies depicting the two bogies and four axles. To establish the carbody critical points, the ride comfort index is calculated in three points deemed relevant in terms of ride comfort evaluation. One of these points is at the carbody center and the other two above the bogies [16, 17].

2. Railway vehicle model and motion equations

The model of the vehicle in figure 1 is a rigid-flexible coupled model, including the flexible carbody, bogies and four axles (figure 1) [11, 19]. The bogies and axles are treated as rigid bodies, since they feature reduced elastic deformations. The vehicle carbody is represented by a free-free equivalent beam, a constant section and uniformly distributed mass, of Euler-Bernoulli type. The parameters of the vehicle model are presented in table 1.

The vehicle is running at constant velocity V on a perfectly rigid track with vertical irregularities. The vertical displacements of the axles $\eta_{1...4}$ are equal with the track vertical irregularities.

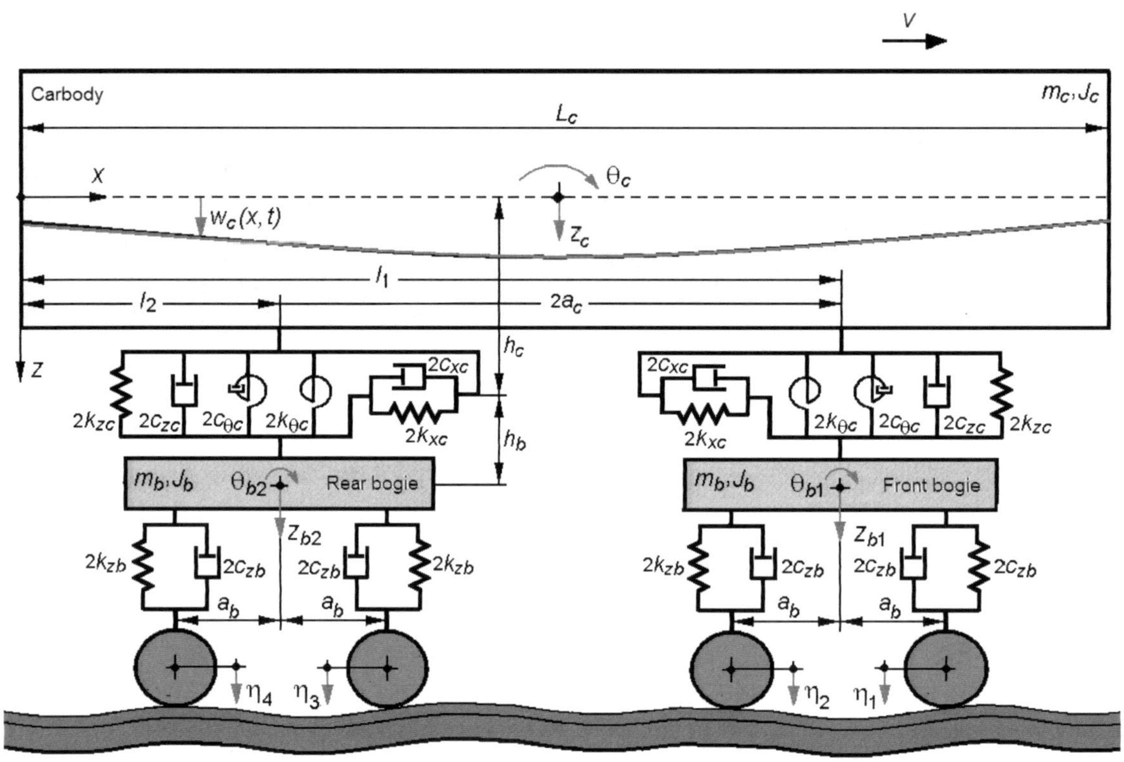

Figure 1. The model of the railway vehicle [11, 19].

The carbody vibrations modes relevant for the ride comfort in a vertical plan are considered: vibration rigid modes - bounce z_c, pitch θ_c, and the first bending mode. The vibration rigid modes of the bogies in a vertical plan are bounce $z_{b1,2}$ and pitch $\theta_{b1,2}$.

The vehicle suspension stages are modelled via Kelvin-Voigt type systems. The secondary suspension related to a bogie has three such systems where two are for translation (vertical and longitudinal) and one for rotation; on the other hand, the primary suspension is modelled through one Kelvin-Voigt system for translation in the vertical direction.

Table 1. The parameters of the vehicle model.

Symbol	Definition	Symbol	Definition
m_c	carbody mass	$2c_{zc}$	vertical damping of the secondary suspension*
m_b	bogie suspended mass	$2k_{zc}$	vertical stiffness of the secondary suspension*
J_c	carbody inertia moment	$2c_{xc}$	longitudinal damping of the secondary suspension*
J_b	bogie inertia moment	$2k_{xc}$	longitudinal stiffness of the secondary suspension*
L_c	carbody length	$2c_{\theta c}$	angular damping of the secondary suspension*
$2a_c$	carbody wheelbase/	$2k_{\theta c}$	angular stiffness of the secondary suspension*
$2a_b$	bogie wheelbase	$2c_{zb}$	vertical damping of the primary suspension**
$l_{1,2}$	position of the carbody on the suspension	$2k_{zb}$	vertical stiffness of the primary suspension**

* per bogie; ** per axle

The motion equation of the carbody vehicle has the general form [11, 19]:

$$EI\frac{\partial^4 w_c(x,t)}{\partial x^4}+\mu I\frac{\partial^5 w_c(x,t)}{\partial x^4 \partial t}+\rho_c\frac{\partial^2 w_c(x,t)}{\partial t^2}=\sum_{i=1}^{2}F_{zci}\delta(x-l_i)+\sum_{i=1}^{2}(M_{ci}-h_cF_{xci})\frac{\mathrm{d}\delta(x-l_i)}{\mathrm{d}x}, \quad (1)$$

where EI is bending modulus (with E - longitudinal modulus of elasticity, and I - inertia moment of the beam's transversal section), μ - structural damping coefficient, and $\rho_c = m_c/L_c$ – beam mass per length unit; $w_c(x, t)$ represents the vertical displacement in a random point of the carbody; $\delta(.)$ is Dirac's delta function; F_{zci}, F_{xci} and M_i stand for forces and moments, respectively, due to the secondary suspension of bogie i (with $i = 1, 2$)

$$F_{zci}=-2c_{zc}\left(\frac{\partial w_c(l_i,t)}{\partial t}-\dot{z}_{bi}\right)-2k_{zc}\left[w_c(l_i,t)-z_{bi}\right]; \quad (2)$$

$$F_{xci}=2c_{xc}\left(h_c\frac{\partial^2 w_c(l_i,t)}{\partial x\partial t}+h_b\dot{\theta}_{bi}\right)+2k_{xc}\left(h_c\frac{\partial w_c(l_i,t)}{\partial x}+h_b\theta_{bi}\right); \quad (3)$$

$$M_{ci}=-2c_{\theta c}\left(\frac{\partial^2 w_c(l_i,t)}{\partial x\partial t}-\dot{\theta}_{bi}\right)-2k_{\theta c}\left[\frac{\partial w_c(l_i,t)}{\partial x}-\theta_{bi}\right]. \quad (4)$$

The vertical displacement of the carbody $w_c(x, t)$ is the result of having overlapped the rigid vibrations modes - bounce and pitch, and the vertical bending [15 - 19],

$$w_c(x,t)=z_c(t)+\left(x-\frac{L_c}{2}\right)\theta_c(t)+X_c(x)T_c(t), \quad (5)$$

where $T_c(t)$ is the coordinate of the carbody bending and $X_c(x)$ stands for the natural function of this vibration mode, described in the equation

$$X_c(x)=\sin\beta x+\sinh\beta x-\frac{(\sin\beta L_c-\sinh\beta L_c)(\cos\beta x+\cosh\beta x)}{\cos\beta L_c-\cosh\beta L_c}, \quad (6)$$

with $$\beta=\sqrt[4]{\omega_c^2\rho_c/(EI)},\ \cos\beta L_c\cosh\beta L_c-1=0, \quad (7)$$

where ω_c is the natural pulsation of the carbody bending.

The application of the modal analysis and the orthogonality property of the eigenfunctions of the carbody bending can help infer the bounce, pitch and vertical bending equations [11, 19],

$$m_c\ddot{z}_c + 2c_{zc}[2\dot{z}_c + 2\varepsilon\dot{T}_2 - (\dot{z}_{b1} + \dot{z}_{b2})] + 2k_{zc}[2z_c + 2\varepsilon T_2 - (z_{b1} + z_{b2})] = 0 \tag{8}$$

$$\begin{aligned} &J_c\ddot{\theta}_c + 2c_{zc}a_c[2a_c\dot{\theta}_c - (\dot{z}_{b1} - \dot{z}_{b2})] + 2k_{zc}a_c[2a_c\theta_c - (z_{b1} - z_{b2})] + \\ &+ 2c_{xc}h_c[2h_c\dot{\theta}_c + h_b(\dot{\theta}_{b1} + \dot{\theta}_{b2})] + 2k_{xc}h_c[2h_c\theta_c + h_b(\theta_{b1} + \theta_{b2})] + \\ &+ 2c_{\theta c}[2\dot{\theta}_c - (\dot{\theta}_{b1} + \dot{\theta}_{b2})] + 2k_{\theta c}[2\theta_c - (\theta_{b1} + \theta_{b2})] = 0; \end{aligned} \tag{9}$$

$$\begin{aligned} &m_{mc}\ddot{T}_c + c_{mc}\dot{T}_c + k_{mc}T_c + 2c_{zc}\varepsilon[2\dot{z}_c + 2\varepsilon\dot{T}_c - (\dot{z}_{b1} + \dot{z}_{b2})] + 2k_{zc}\varepsilon[2z_c + 2\varepsilon T_c - (z_{b1} + z_{b2})] + \\ &+ 2c_{xc}h_c\lambda[2h_c\lambda\dot{T}_c + h_b(\dot{\theta}_{b1} - \dot{\theta}_{b2})] + 2k_{xc}h_c\lambda[2h_c\lambda T_c + h_{b2}(\theta_{b1} - \theta_{b2})] + \\ &+ 2c_{\theta c}\lambda[2\lambda\dot{T}_c - (\dot{\theta}_{b1} - \dot{\theta}_{b2})] + 2k_{\theta c}\lambda[2\lambda T_c - (\theta_{b1} - \theta_{b2})] = 0, \end{aligned} \tag{10}$$

where k_{mc} is the carbody modal stiffness, c_{mc} - carbody modal damping and m_{mc} - carbody modal mass

$$k_{mc} = EI\int_0^L \left(\frac{\mathrm{d}^2X_c}{\mathrm{d}x^2}\right)^2 \mathrm{d}x\,;\; c_{mc} = \mu I\int_0^L \left(\frac{\mathrm{d}^2X_c}{\mathrm{d}x^2}\right)^2 \mathrm{d}x\,;\; m_{mc} = \rho_c\int_0^L X_c^2\mathrm{d}x\,. \tag{11}$$

The notations ε and λ have been introduced, based on the symmetry properties of the eigenfunction $X_c(x)$,

$$X_c(l_1) = X_c(l_2) = \varepsilon\,; \tag{12}$$

$$\frac{\mathrm{d}X_c(l_1)}{\mathrm{d}x} = -\frac{\mathrm{d}X_c(l_2)}{\mathrm{d}x} = \lambda\,. \tag{13}$$

The equations describing the bounce and pitch motions of the bogies are

$$m_b\ddot{z}_{b1} = \sum_{i=1}^{2} F_{zbi} - F_{zc1},\; m_b\ddot{z}_{b2} = \sum_{i=3}^{4} F_{zbi} - F_{zc2}, \tag{14}$$

$$J_b\ddot{\theta}_{b1} = a_b\sum_{i=1}^{2} \pm F_{zbi} - h_bF_{xc1},\; J_b\ddot{\theta}_{b2} = a_b\sum_{i=3}^{4} \pm F_{zbi} - h_bF_{xc2}, \tag{15}$$

where $F_{zb1...4}$ stands for the forces due to the primary suspension, given in the relations below

$$F_{zb1,2} = -2c_{zb}(\dot{z}_{b1} + a_b\dot{\theta}_{b1} - \dot{\eta}_{1,2}) - 2k_{zb}(z_{b1} + a_b\theta_{b1} - \eta_{1,2})\,; \tag{16}$$

$$F_{zb3,4} = -2c_{zb}(\dot{z}_{b2} - a_b\dot{\theta}_{b2} - \dot{\eta}_{3,4}) - 2k_{zb}(z_{b2} - a_b\theta_{b2} - \eta_{3,4})\,; \tag{17}$$

The motion equations of the bogies write as such [14, 19]:

$$\begin{aligned} &m_b\ddot{z}_{b1} + 2c_{zb}[2\dot{z}_{b1} - (\dot{\eta}_1 + \dot{\eta}_2)] + 2k_{zb}[2z_{b1} - (\eta_1 + \eta_2)] + \\ &+ 2c_{zc}(\dot{z}_{b1} - \dot{z}_c - a_c\dot{\theta}_c - \varepsilon\dot{T}_c) + 2k_{zc}(z_{b1} - z_c - a_c\theta_c - \varepsilon T_c) = 0; \end{aligned} \tag{18}$$

$$\begin{aligned} &J_b\ddot{\theta}_{b1} + 2c_{zb}a_b[2a_b\dot{\theta}_{b1} - (\dot{\eta}_1 - \dot{\eta}_2)] + 2k_{zb}a_b[2a_b\theta_{b1} - (\eta_1 - \eta_2)] + \\ &+ 2c_{xc}h_b[h_b\dot{\theta}_{b1} + h_c(\dot{\theta}_c + \varepsilon\dot{T}_c)] + 2k_{xc}h_b[h_b\theta_{b1} + h_c(\theta_c + \varepsilon T_c)] = 0 \end{aligned} \tag{19}$$

$$m_b\ddot{z}_{b2}+2c_{zb}[2\dot{z}_{b2}-(\dot{\eta}_3+\dot{\eta}_4)]+2k_{zb}[2z_{b2}-(\eta_3+\eta_4)]+ \\ +2c_{zc}(\dot{z}_{b2}-\dot{z}_c+a_c\dot{\theta}_c-\varepsilon\dot{T}_c)+2k_{zc}(z_{b2}-z_c+a_c\theta_c-\varepsilon T_c)=0 \tag{20}$$

$$J_b\ddot{\theta}_{b2}+2c_{zb}a_b[2a_b\dot{\theta}_{b2}-(\dot{\eta}_3-\dot{\eta}_4)]+2k_{zb}a_b[2a_b\theta_{b2}-(\eta_3-\eta_4)]+ \\ +2c_{xc}h_b[h_b\dot{\theta}_{b2}+h_c(\dot{\theta}_c-\varepsilon\dot{T}_c)]+2k_{xc}h_b[h_b\theta_{b2}+h_c(\theta_c-\varepsilon T_c)]=0. \tag{21}$$

The motion equations for the carbody and bogies (8-10 and 18-21) make up a 7-equation system with ordinary derivatives, which can be solved numerically using MATLAB code. The system can be matrix-like written

$$\mathbf{M}\ddot{\mathbf{p}}+\mathbf{C}\dot{\mathbf{p}}+\mathbf{K}\mathbf{p}=\mathbf{P}\dot{\boldsymbol{\eta}}+\mathbf{R}\boldsymbol{\eta}, \tag{22}$$

where **M**, **C** and **K** are the inertia, damping and stiffness matrices, $\mathbf{p}=[z_c\ \ \theta_c\ \ T_c\ \ z_{b1}\ \ \theta_{b1}\ \ z_{b2}\ \ \theta_{b2}]^{\mathrm{T}}$ represents the vector of the displacements' coordinates and $\boldsymbol{\eta}$ is the vector of the heterogeneous terms.

3. Calculation of the ride comfort index

To evaluate the ride comfort in the vertical vibrations, the partial ride comfort index is calculated [16]

$$N_{MV}=6a^{W_{ab}}, \tag{23}$$

where

$$a(x)=\sqrt{\frac{1}{\pi}\int_0^\infty G_{ac}(\mathrm{x},\omega)\mathrm{d}\omega}, \tag{24}$$

is the root mean square of the vertical acceleration and $W_{ab}=W_a{\cdot}W_b$ represents the weight filter of the accelerations [16, 17]. The filter W_a is a pass band filter and the weighting filter W_b takes into account the high human sensitivity to the vertical vibrations. The root mean square of the vertical acceleration is calculated according to the power spectral density of the carbody acceleration

$$G_{ac}(x,\omega)=G(\omega)\omega^4\left|\overline{H}_{z_c}(\omega)+\left(\frac{L_c}{2}-x\right)\overline{H}_{\theta_c}(\omega)+X_c(x)\overline{H}_{T_c}(\omega)\right|^2, \tag{25}$$

where $\overline{H}_{z_c}(\omega)$, $\overline{H}_{\theta_c}(\omega)$, $\overline{H}_{T_c}(\omega)$ are the frequency response functions corresponding to the bounce and pitch and to the vertical bending of the carbody, and $G(\omega)$ is the power spectral density of the vertical track irregularities,

$$G(\omega)=\frac{A\Omega_c^2V^3}{[\omega^2+(V\Omega_c)^2][\omega^2+(V\Omega_r)^2]}, \tag{26}$$

where Ω_c = 0.8246 rad/m, Ω_r = 0.0206 rad/m, and A = 4.032·10^{-7} radm - for a high level quality track, A = 1.080·10^{-6} radm - for a low level quality track [18].

When adopting the hypothesis that the vertical accelerations have a Gaussian distribution with the null mean value, we will have the following relation to calculate the partial ride comfort index [19]

$$N_{MV}(x)=6\Phi^{-1}(0.95)\sqrt{\frac{1}{\pi}\int_0^\infty G_{ac}(x,\omega)\left|H_{ab}(\omega)\right|^2\mathrm{d}\omega}, \tag{27}$$

where $\Phi^{-1}(0.95)$ represents the quantile of the standard Gaussian distribution with the probability of 95%, and $H_{ab}(\omega)=H_a(\omega){\cdot}H_b(\omega)$, where $H_a(\omega)$ and $H_b(\omega)$ are the transfer functions corresponding to the filters W_a and W_b [16, 17],

$$H_a(s)=\frac{s^2(2\pi f_2)^2}{\left[s^2+\frac{2\pi f_1}{Q_1}s+(2\pi f_1)^2\right]\left[s^2+\frac{2\pi f_2}{Q_1}s+(2\pi f_2)^2\right]} \tag{28}$$

$$H_b(s)=\frac{(s+2\pi f_3)\cdot\left[s^2+\frac{2\pi f_5}{Q_3}s+(2\pi f_5)^2\right]2\pi K f_4^2 f_6^2}{\left[s^2+\frac{2\pi f_4}{Q_2}s+(2\pi f_4)^2\right]\left[s^2+\frac{2\pi f_6}{Q_4}s+(2\pi f_6)^2\right]f_3 f_5^2} \tag{29}$$

where $s = \mathrm{i}\omega$ (with $\mathrm{i}^2 = -1$), $f_1 = 0.4$ Hz, $f_2 = 100$ Hz, $f_3 = 16$ Hz, $f_4 = 16$ Hz, $f_5 = 2.5$ Hz, $f_6 = 4$ Hz, $Q_1 = 0.71$, $Q_2 = 0.63$, $Q_4 = 0.8$ and $K = 0.4$.

4. The results of the numerical simulations

This section features the numerical simulations results relating to the ride comfort index along the carbody and the position of the critical points in terms of the ride comfort, in correlation with the speed, carbody flexibility and suspension damping. The numerical simulation applications are developed in MATLAB numerical computing environment.

To examine the position of the critical points of the ride comfort, three important carbody points are taken into account, from the viewpoint of ride comfort evaluation – one at the carbody center and two points above the bogies.

The damping ratios corresponding to the two levels of suspension will be brought in, as below

$$\zeta_c=\frac{4c_{zc}}{2\sqrt{4k_{zc}m_c}};\ \zeta_b=\frac{4c_{zb}}{2\sqrt{4k_{zb}m_b}}. \tag{30}$$

The numerical simulation parameters (reference parameters of vehicle) are presented in table 2. According to these parameters, the natural frequencies of the carbody bending f_b is 8 Hz and the damping ratios of suspension are $\zeta_c = 0.12$ – for the secondary suspension and $\zeta_b = 0.22$ – for the primary suspension.

Table 2. The parameters of the vehicle numerical model [11, 19].

$m_c = 34.0\cdot10^3$ kg	$L_c = 26$ m	$k_{zc} = 0.60$ MN/m
$m_b = 3.20\cdot10^3$ kg	$h_c = 1.30$ m	$k_{xc} = 2.00$MN/m
$m_{mc} = 35.2\cdot10^3$ kg	$h_b = 0.20$ m	$k_{\theta c} =$ 128 kN/m
$k_{mc} = 89.0$ MN/m	$2a_c = 19.0$ m	$c_{zc} = 17.15$ kNs/m
$c_{mc} = 53.1$ kNm/s	$2a_b = 2.56$ m	$c_{xc} = 25.0$ kNs/m
$J_c = 1.96\cdot10^6$ kg·m^2	$l_1 = 3.5$ m	$c_{\theta c} = 1$ kNm
$J_b = 2.05\cdot10^3$ kg·m^2	$l_2 = 22.5$ m	$k_{zb} = 1.10$ MN/m
$EI = 3.02\cdot10^9$Nm^2	$A = 1.080\cdot10^{-6}$ radm	$c_{zb} = 13.05$ kNs/m

Figure 2 shows the root mean square (rms) of vertical acceleration along the vehicle carbody for more velocities. First, the level of carbody vibrations is noticed to increase with the velocity. Second, the behaviour of vibrations along the carbody is not symmetrical compared to its center. In general, the root mean square of vertical acceleration is lower at the center and raises at the carbody ends.

Figure 3 shows that the ride comfort index increases with the velocity, which means a worse ride comfort. For velocities up to 200 km/h, the ride comfort index is seen to be lower at the carbody center and higher towards its ends. Should velocity goes up (V = 250 km/h or V = 300 km/h), a significant increase of the ride comfort index occurs at the carbody center. Also, the ride comfort index has an asymmetrical distribution along the carbody versus its center.

In diagram (b), the critical points of the ride comfort are located above one of the two bogies, and their position changes upon velocity. At velocity of 100 km/h and 150 km/h, the critical point of the ride comfort is placed above the front bogie, while this point moves above the rear bogie for higher velocities. It is important to notice that the ride comfort index at the carbody center is higher than this index above the front bogie, for velocities of 250 km/h and 300 km/h.

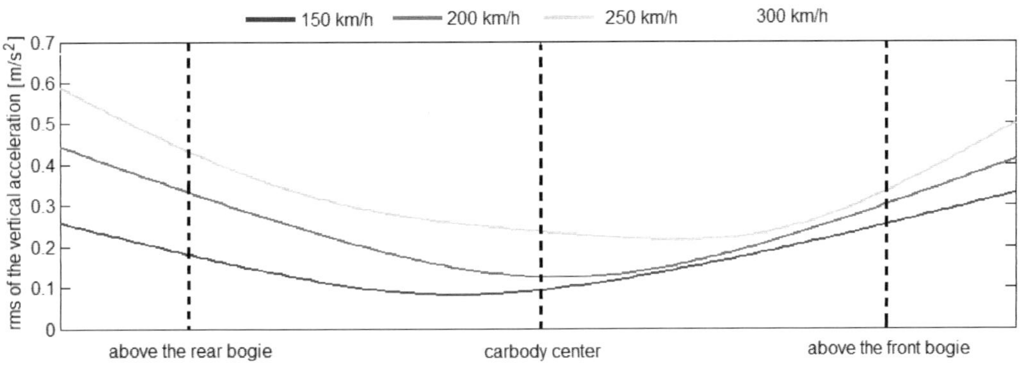

Figure 2. Root mean square of vertical acceleration along the carbody.

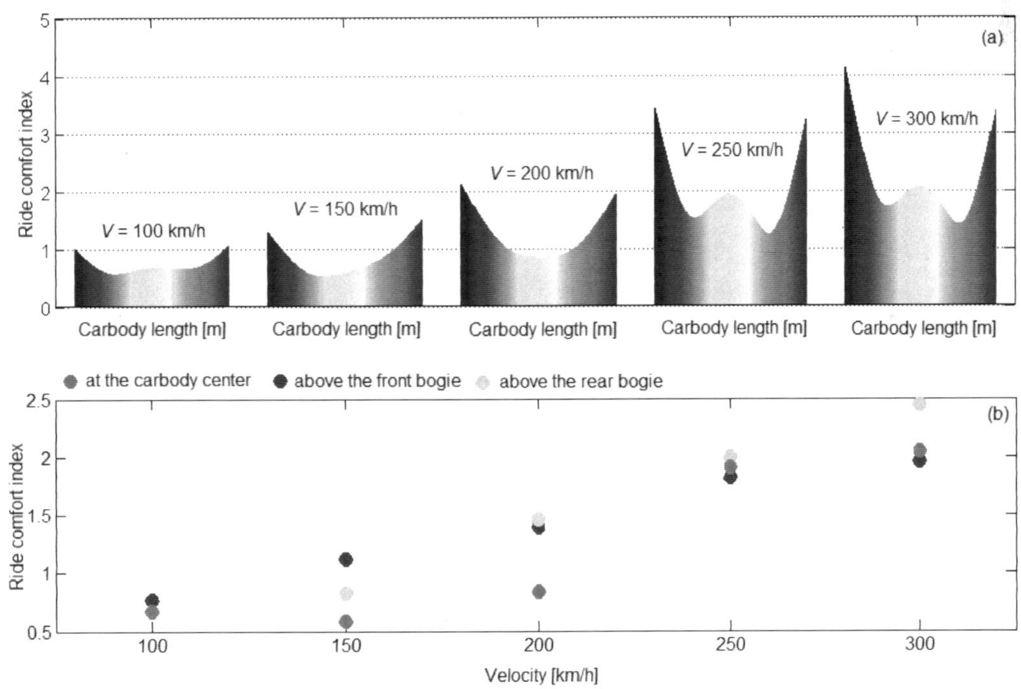

Figure 3. Ride comfort index – influence of the velocity: (a) along the carbody; (b) at the carbody center and above the bogies.

The ride comfort is strongly influenced by the carbody flexibility, as shown in figure 4. A higher bending modulus means a lower comfort index along the entire carbody. A decrease in the comfort index is more visible at the carbody center. Nevertheless, significant improvement of the ride comfort is not acquired beyond a certain value of the bending modulus ($EI = 4.71 \cdot 10^9$ Nm^2). For all the cases analyzed, the critical point of the ride comfort is found above the rear bogie (see diagram (b)).

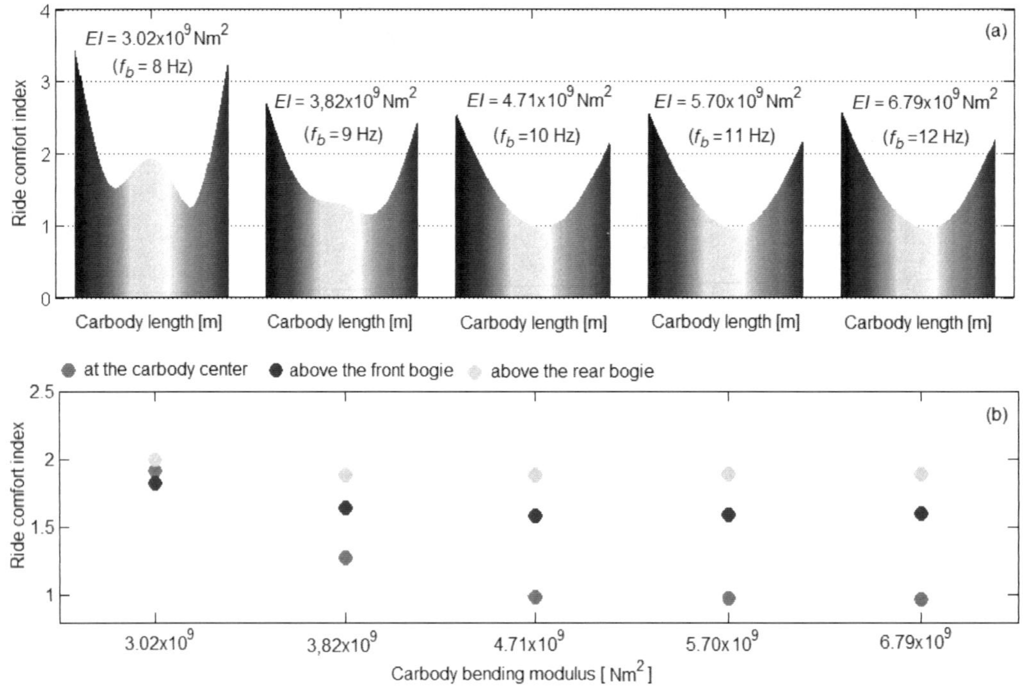

Figure 4. Ride comfort index of 250 km/h – influence of the carbody bending modulus: (a) along the carbody at velocity; (b) at the carbody center and above the bogies.

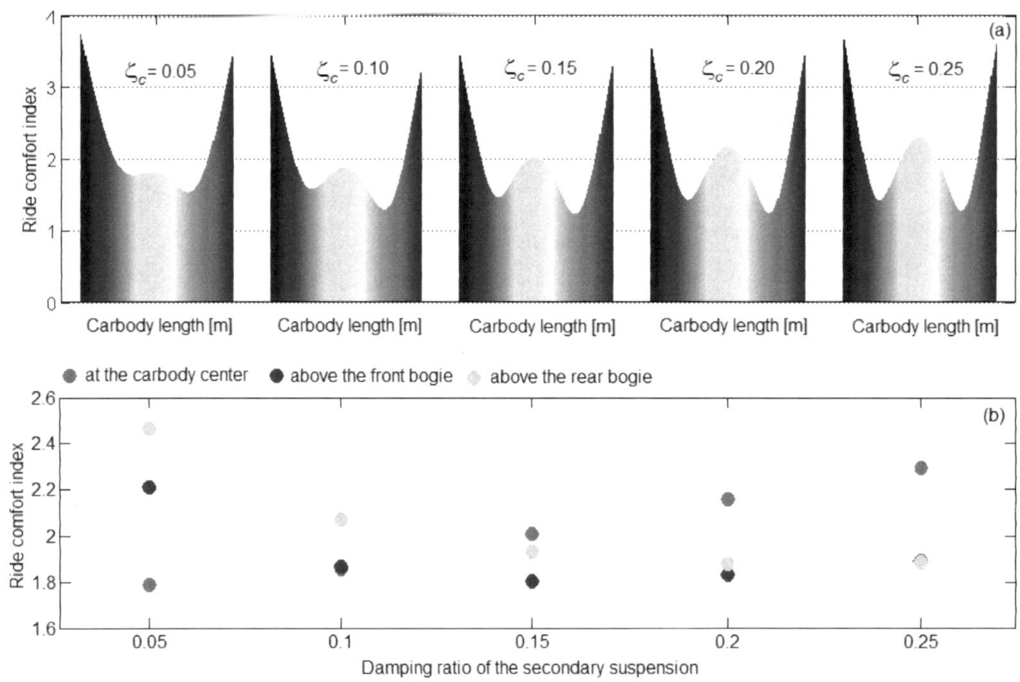

Figure 5. Ride comfort index of 250 km/h – influence of the on the damping ratio of the secondary suspension: (a) along the carbody at velocity; (b) at the carbody center and above the bogies.

IOP Conf. Series: Materials Science and Engineering **682** (2019) 012004 doi:10.1088/1757-899X/682/1/012004

Figure 5 shows the influence that the damping ratio of the secondary suspension has upon the ride comfort index. The increase of the damping ratio of the secondary suspension has contrary effects upon the ride comfort. While the ride comfort becomes better at the carbody ends this it degrades considerably at the carbody center. Such observations are much clearer in diagram (b). The same diagram underlies the determination of the critical points in the ride comfort. For small values of the damping ratio of the secondary suspension ($\zeta_c \leq 0.1$), the critical point in the ride comfort is noted to be above the rear bogie. When the damping ratio of the secondary suspension raises, the critical point in the ride comfort moves to the carbody center.

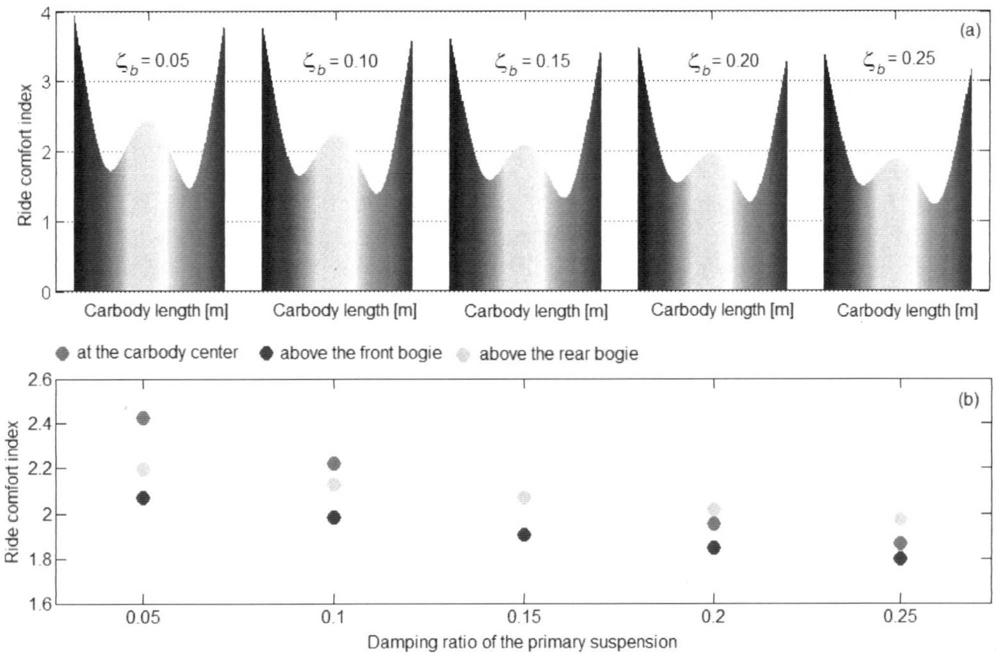

Figure 6. Ride comfort index of 250 km/h – influence on the damping ratio of the primary suspension: (a) along the carbody at velocity; (b) at the carbody center and above the bogies.

The influence of the damping ratio of the primary suspension upon the ride comfort index is shown in the diagrams of figure 6. An increased damping ratio of the primary suspension results into a better ride comfort along the entire carbody, yet the most important results are derived at the carbody center. The damping ratio of the primary suspension also affects the position of the critical points in the ride comfort, as seen in diagram (b). For $\zeta_b \leq 0.1$, the critical point in the ride comfort is at the carbody center; this point moves above the rear bogie when the damping ratio of the primary suspension has higher values.

5. Conclusions

The paper herein submits an analysis of the ride comfort at vertical vibrations along the railway vehicle carbody and the position of the critical points in the ride comfort. To this end, the results to the ride comfort index derived from the numerical simulations are used. Numerical simulations are developed by a rigid-flexible coupled model type of vehicle, where the carbody vibration modes relevant for the ride comfort in a vertical plan are taken into account - bounce, pitch and the first bending modulus.

The observations made have pointed out at the fact that the ride comfort index rises along with velocity, plus the asymmetrical distribution of the ride comfort index reported to the carbody center. The ride comfort index is generally lower at the carbody center and rises towards its ends. Under certain circumstances – flexible carbody, high velocities, high damping ratio of the secondary suspension or low damping ratio of the primary suspension, the ride comfort index rises a great deal at the carbody center.

Considering three important points of the carbody from the ride comfort evaluation perspective – a point located at the carbody center and two points above the bogies, it turned out that any of these points can represent the critical point of the carbody in terms of ride comfort. The position of the critical point of the ride comfort changes depending on the vehicle velocity, carbody flexibility and the suspension damping. The points above the bogies are normally critical points of the ride comfort. However, the critical point in the ride comfort is placed at the carbody center, for a high damping ration of the secondary suspension or a low damping ratio of the primary suspension.

6. References

[1] Förstberg J 2000 *Ride comfort and motion sickness in tilting trains* - Doctoral thesis (Stockholm: Railway Technology Department of Vehicle Engineering, Royal Institute of Technology)
[2] Mansfield N J 2005 *Human response to vibration* (CRC Press LLC)
[3] Griffin M J 1990 *Handbook of Human Vibration* (London: Academic Press)
[4] Richards G L 1980 *On the psychology of passenger comfort* in *Human factors in transport* research, vol. 2 (New York: Academic Press)
[5] Suzuki H 1998 Effects of the range and frequency of vibrations on the momentary riding comfort evaluation of a railway vehicle *Jpn. Psychol. Res.* **40(3)** 156-165
[6] Kardas-Cinal E 2010 Ride comfort for various passenger positions in a railway vehicle – simulation study *The Archives of Transport* **22(2)**189-199
[7] Kim Y G, Choi S, Kim S W, Kim Y M and Park T W 2009 An experimental study on the ride comfort of the Korean high-speed train *Exp. Tech.* **33(6)** 30-37
[8] Kim Y G, Kwon H B, Kim S W, Park C K and Park T W 2003 Correlation of ride comfort evaluation methods for railway vehicles *Proc. of the Institution of Mechanical Engineers, Part F: J. of Rail and Rapid Transit* **217** 3-88
[9] Wei X, Liu Y, Sun Q, Jia L and Wang Y 2013 Rail vehicle ride comfort prediction based on bogie acceleration measurements *Proc. Of 25th Chinese Control and Decision Conference* (CCDC) 3810-3815
[10] Iwnicki S 2006 *Handbook of railway vehicle dynamics* (Boca Raton: Taylor & Francis Group)
[11] Dumitriu M 2014 On the critical points of vertical vibration in a railway vehicle *Arch. Mech. Eng.* **LXI(4)** 115-140
[12] Gong D, Zhou J and Sun W 2017 Passive control of railway vehicle car body flexural vibration by means of under frame dampers *J. Mech. Sci. Technol.* **31(2)** 555-564
[13] Dumitriu M 2018 Study on improving the ride comfort in railway vehicles using anti-bending dampers *Appl. Mech. Mater.* **880** 207-212
[14] Dumitriu M 2017 A new approach to reducing the carbody vertical bending vibration of railway vehicles *Veh. Syst. Dyn.* 55(11) 1787-1806
[15] Dumitriu M and Cruceanu 2017 Influences of carbody vertical flexibility on ride comfort of railway vehicles *Arch. Mech. Eng.* **64(2)** 119-238
[16] EN 12299:1997 *Railway applications ride comfort for passengers measurement and evaluation*
[17] UIC 513 R:1994 *Guidelines for evaluating passenger comfort in relation to vibration in railway vehicle, International Union of Railways*
[18] C 116 Interaction between vehicles and track, RP 1, Power spectral density of track irregularities, Part 1: Definitions, conventions and available data, Utrecht, 1971
[19] Dumitriu M 2017 Ride comfort enhancement in railway vehicle by the reduction of the carbody structural flexural vibration *IOP Conf. Ser. Mater. Sci. Eng.* **227** 012042

IOP Conf. Series: Materials Science and Engineering **682** (2019) 012005 doi:10.1088/1757-899X/682/1/012005

Analysis on the measurement results of the precision grades and their influence on the performance of the hydraulic pumps with spur gears

I G Ghionea[1], A L Ghionea[2], S Ćuković[3] and M I Pleşa[4]

[1] Manufacturing Engineering Department, University Politehnica of Bucharest, Faculty of Industrial Engineering and Robotics, 313 Splaiul Independentei, Bucharest 060042, Romania

[2] Machine and Production Systems Department, University Politehnica of Bucharest, Faculty of Industrial Engineering and Robotics, 313 Splaiul Independentei, Bucharest 060042, Romania

[3] Institute for Information Technologies, University of Kragujevac, Jovana Cvijica street, Kragujevac 34000, Serbia

[4] Doctoral School of Faculty of Industrial Engineering and Robotics, University Politehnica of Bucharest, 313 Splaiul Independentei, Bucharest 060042, Romania

ionut.ghionea@imst.pub.ro

Abstract. The paper presents results of experimental measurements on the main precision grades for a cylindrical spur gearing, from the assembly of a hydraulic pump integrated in a hydrostatic driving system. For the components of this assembly, the main stages of the design, material characteristics, constructive, precision and functional parameters that define the flow, the pressure of the hydraulic environment, the driving speed are briefly presented. The analysis was performed based on the data obtained by measuring the gears' deviations and tolerances. Among these, the most important data for complying the optimal requirements of the pump operating criteria were retained and analysed. The analysis of the measured results highlighted the main causes and effects of the deviations found. The presented results show the problem's complexity and are important in evaluating the performances that characterize these types of pumps, with wide use in many fields of hydraulic actuators.

1. Introduction

Hydrostatic systems (of actuation, lubrication, cooling, etc.) have a wide use, corresponding to the technical and performance data required for different industrial fields, such as: construction of machine-tools, automotive, transport equipment and installations, etc. The basic component of these systems is the hydraulic pump (generator), that is manufactured in many constructive and functional variants for a wide range of flow rates and pressures [1].

Hydraulic pumps are used in most of the hydraulic power transmission systems. These pumps convert mechanical energy into hydraulic energy, that is a combination of pressure and flow using a specific constructive solution. Different types of hydraulic pumps can be integrated into a wide variety of industrial hydraulic machines, being used on machine-tools, presses (for shearing, stamping, bending), excavators, cranes, loaders, tractors, forestry equipment, mining machinery, injection moulding machines etc.

IOP Conf. Series: Materials Science and Engineering **682** (2019) 012005 doi:10.1088/1757-899X/682/1/012005

There are two main groups of pumps: with a constant flow and with an adjustable flow. The main characteristics of the pumps refer to: constructive typology, overall dimensions, flow rate, working pressure, driving speed range and torque, efficiency, acquisition costs of some components, operation and maintenance, reliability, etc. The technical performances are determined by the following requirements, namely: the maximum driving speed and pressure level, the geometric volume (capacity), reduced pulses of the flow, increased reliability, low costs per unit of power [2]. Flows, discharge pressures and driving speeds are specific parameters, with values within wide limits, corresponding to the requirements of the installation in which the pump operates. Thus, it is appropriate to manufacture the pumps in standardized series [3], on a catalogue basis.

The spur gear pump assembly, analysed in this paper, was chosen from a constructive-functional wide range of pumps with different uses. This studied pump is new/unused (only ran for testing on the stand) and it is part of a manufacturing lot from a Romanian producer. The analysis presented aims to identify the fulfilment of the manufacturing and precision requirements imposed by the use.

The pump respects a series of requirements regarding the operation of the spur gearing in conditions of variable loads, of low vibration and noise, durability, safety in operation and optimum efficiency [4, 5]. These determine specific conditions of design, manufacture and control [6] of the two gears, their shafts and bearings blocks, of the functional and assembly surfaces of the pump components.

2. Stages of design and manufacture of the spur gear pump

Generally, the development of new products and/or the improvement of existing ones are important research and design activities in any company in the field of industrial manufacturing. Thus, in these activities there are certain risks and unexpected events that may reduce the technical characteristics. The causes can be numerous: deviations from the technical documentation, errors of processing or of measurement, defects of material, wear of the cutting tools, different elastic deformations, vibrations in the processes of manufacturing, quality and viscosity of the hydraulic fluid, impurities, variations of temperature, etc. All these affect the operating behaviour and decrease the pump performance [7].

The improvement of the constructive variant and the choice of materials require the simulation of the pump design (figure 1) in the stages of: computer-aided parts and technological design, creation of virtual 3D models, identification of several variants for the same product (using parameterized design) [8, 9], correct choice of manufacturing technologies, control methodologies and assembly of components, analysis of functioning behaviour. Appropriately, the manufacturing techniques are optimized and the costs and delivery times are established.

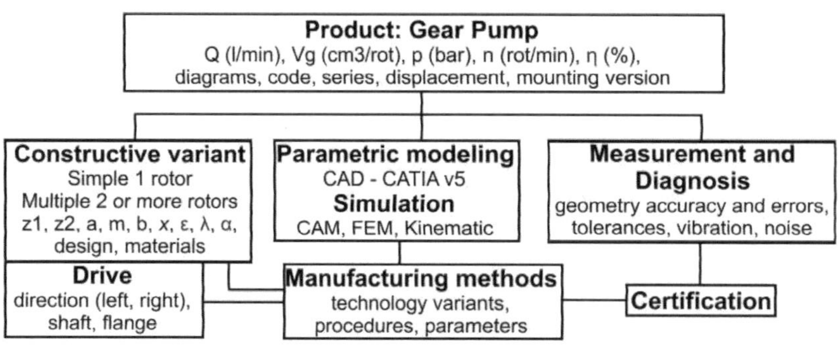

Figure 1. Design stages of the gear pump product

The main data that are involved in the parts and technological design stages are: geometric and operating parameters of the pump (Q_p - flow rate, V_g - geometric volume/displacement, p_p - working pressure, volume per one revolution, n_0 - speed, η - efficiency, other conditions and constraints, like: how is the end of the driving shaft, the shape and dimensions of the clamping flange, etc.). There are considered some constructive variants (with one, two or more pairs of gears, number of teeth, gear module, gear width, pressure angle). Generally, the pump is driven by a constant speed electric motor.

An important step in improving the performances of the pump is the optimization of the teeth flanks generation by: modification of the profile, flanking, rounding etc. [10] Thus, by parametric design, the 3D model of the gearing and of the pump assembly is obtained (figure 2). The applied methodology is based on numerous and complex parametric relations [1, 11] between certain dimensions that define shapes and positions of the pump components.

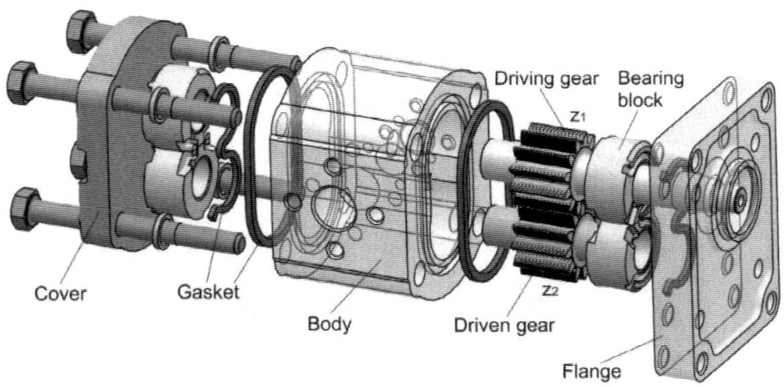

Figure 2. Parametric 3D model of the gear pump

The pump flow is provided by the size and number of gaps between the teeth of the cylindrical gear ($z_1 = z_2$) and its driving speed n_0 (rev/min). The hydraulic fluid is aspirated at the atmospheric pressure p_0 (bar) and discharged at a flow rate Q_p (l/min) and pressure p_p (bar) through the inlet, respectively, the outlet apertures of the pump body. The flow is determined by the geometric elements of the gears' teeth [5] supported by the bearings.

The 3D model of the pump in figure 2 was obtained using the following parameters: number of teeth $z_1 = z_2 = 12$, centre distance $A = 31.4$ mm, module $m = 2.54$ mm (normal specific displacements $x_n = 0.2$ mm pinion, $x_n = 0.199$ mm gear and frontal specific displacement $x_f = 0.2$ mm), gears width $b = 18.3$ mm, pitch circle diameter $d_d = 30.48$ mm, outside circle diameter $d_a = 37.413$ mm, rolling circle diameter $d_r = 31.4$ mm, base circle diameter $d_b = 28.642$ mm, pressure angle $\alpha = 20^0$, teeth height $h = 6.205$ mm, addendum $h_a = 1.261 \times m = 3.203$ mm, normal gear pitch p_n, normal arc of dividing the tooth/gap s_n/e_n, frontal contact ratio $\varepsilon = 1.494$, constant chord $s_c = 3.85$ mm. The parameter values are based on calculation methods [1, 7] and standards in the field [11]. These parameters should be found in the measurement process of the manufactured pump. Analysis of these data shows that the involute tooth gear is corrected.

High speeds, over $n_0 = 1500 \ldots 2000$ rev / min, lead to appearance of the cavity process, which reduces the pump flow, efficiency and also to gas occurrence in the hydraulic fluid.

The market demands and pump producers' concerns prove interest for the parametric design of pumps' series. Today's computer aided design techniques (CAD) offer many possibilities to edit the parameters in order to obtain other custom versions of the designed pumps. The validation of the pumps series is possible and is done by modern and complex simulations, such as: kinematics, FEM, CAM etc.

3. Measured data and results

Figure 3 presents the constructive solution of the pump type PRD2-223S, series 337 [3], having the geometric volume $V_g = 11$ cm^3 / rev, the suction pressure $p_a = 0.7 \ldots 3$ bar, the nominal pressure $p_n = 250$ bar, the maximum allowable pressure $p_{max} = 280$ bar, minimum speed $n_{min} = 500 \ldots 1200$ rev / min, maximum speed $n_{max} = 3000 \ldots 3500$ rev / min.

IOP Conf. Series: Materials Science and Engineering **682** (2019) 012005 doi:10.1088/1757-899X/682/1/012005

The pump for which measurements were made (gears and shafts, body, bearing blocks) has an external cylindrical spur gearing with an involute profile [1, 10], the body is provided with two apertures: inlet for suction the fluid at the atmospheric pressure p_0 and outlet to discharge it in the hydraulic installation at the nominal pressure p_n. The bearing blocks have also the role of closing the gearing in the pump's body and all are assembled by the pump's cover and flange. The leakage of fluid outside the pump is prevented by rubber gaskets, according to figures 2 and 3.

Figure 3. The analysed pump in disassembled representation

A metallographic analysis was performed on some components of the pump, which showed up the following materials: pump body (AlSi2MgTi), bearing blocks (AlSi6Cu4), flange and cover (EN-GJL-HB195), gears and their shafts (316 stainless steel).

In the research of the project [12], we have observed and studied, among others: the influence of the errors of manufacturing the main parts and of their assembly in the pump, the execution quality of teeth flank surfaces on the constructive and functional performances.

This paper is a continuation of the project and deals with the measurement of certain important parameters of the gears. The studies revealed that the errors and deviations of the gears elements and bearings occur during the different manufacturing stages of the body, gears, teeth flanks, shafts bearing blocks, when mounting the gearing in the pump assembly and under operating conditions. To determine the causes and sources of errors, specific steps, rules and methodologies presented in specialized standards [13, 14], workbooks and scientific papers [1, 7, 9] were applied.

In many analyses it was identified which of the sources have a higher relevance: the machine-tool, the cutting tools, instruments and control devices. These are basic components of the technological system. Also, it was necessary to analyse the process and manufacturing technology and, implicitly, the values of the process parameters.

For this pump, the research in the project was carried out, mainly, in three directions, regarding:

The processing precision of the spur gears, corresponding to their type-dimensions, that is defined by the precision grade, the roughness of the flanks and the clearance between the flanks.

Precision indices regarding the names and their notation, as well the sizes of the limit deviations and the specific tolerances, are established by standards [15]. These indicate individual errors and cumulative errors.

Precision criteria. In defining them, we consider the main areas of gears use and the grouping of the precision grades indices. According to the three corresponding precision criteria: kinematic, smooth running and contact between teeth [14], measurements were made using the Klingelnberg P150 centre. For each criterion there are established indices and groups of precision indices.

IOP Conf. Series: Materials Science and Engineering **682** (2019) 012005 doi:10.1088/1757-899X/682/1/012005

Thus, for this type of hydraulic pumps, the following deviation values are recommended [7], being important and normal in the assembly process and during the pump operation: the front clearance between the front surfaces of the gears and bearings: 10 ... 50 μm, the tolerance of the distance between the gears axes: max. 0.02 mm, the eccentricity between the gears outside diameters and the bores diameters of the body: max. 0.01 mm, radial clearance of the pump > 0.0015 × outer diameter of the gears, parallelism of the front surfaces should not exceed a tolerance of 0.01 mm.

The measurements were made for each gear in the transverse plane (figure 4), on the left and on the right flanks of the 12 teeth.

Figure 4. Control of the profile and of the teeth flank line

Diagrams in figure 5 (for the driving gear/pinion z_1) present results of these measurements on flanks and the value of the tolerance prescribed for the precision grade 5.

According to figure, the notations are: F_p - total cumulative pitch deviation (difference between the most positive and the most negative index value for all teeth), f_p –single pitch deviation (difference between two adjacent teeth index values +, -), $f_{p\ max}$ – maximum single pitch deviation (maximum difference between two adjacent teeth index values +, -), $f_{u\ max}$ – maximum difference between adjacent pitches (maximum difference between actual dimensions of two successive right or left flank transverse pitches), R_p – range of pitch error, pitch variation (difference between the largest and the smallest actual size of the transverse pitches of a given flank), $F_{p\ z/8}$ – total cumulative pitch deviation for eight teeth (maximum difference in cumulative deviations of actual values and those nominal for eight teeth), F_r – runout error (radial position difference of a probe contacting all teeth at measuring diameter – pitch diameter, it combines tooth eccentricity in relation to the datum axis and the tooth spacing error), R_s – variation of tooth thickness (difference between the largest and the smallest tooth thickness of a gear) [14, 15].

The results of the measurements presented in the diagrams (figure 5) for the deviations f_p and F_p do not correspond (f_p = 21.7 μm and f_p = 27.5 μm, the left flank, respectively, the right flank) because the left flank has an addendum and a dedendum relief/flanking. The maximum value prescribed in the standard [15] for f_p is 5 μm in the precision grade 5, but the values resulting from the measurements correspond to the precision grade 10. Similarly, the values measured for the deviation F_p (F_p = 40.9 μm and F_p = 52.1 μm, the left flank, respectively, the right flank) do not correspond to the prescribed value of 16 μm according to the precision grade 5, being (as values) in the precision grade 9.

There were determined total cumulative pitch deviations for eight teeth $F_{p\ z/8}$ = 40.2 μm and $F_{p\ z/8}$ = 44.1 μm for the left and right flank, respectively, and correspond to precision grade 10. DIN 3962 [15] provides a maximum value of 10 μm for these indices, according to the precision grade 5.

The runout error F_r defines the runout of the pitch circle, it is the error in radial position of the teeth. Most often it is measured by indicating the position of a pin or a ball inserted in each tooth space around the gear and taking the largest difference. This error causes a number of problems: noise and wear of gears and bearing blocks. Its source is most often insufficient accuracy and ruggedness of the driving shaft and of the tooling system. In the size of the runout error, the eccentricity of the gear influences the general shape of the gear kinematic error diagram. This is used to determine several errors and individual deviations.

Spur gear division		
Name: Driving gear/Pinion	No. of teeth: 12	Pressure angle: 20°00'00''
Drawing no. Pinion z_1=12	Module m=2.54 mm	Flank line inclination angle: 0°0'0''
Order no. 01	Condition: final	Place of measurement: Neptun

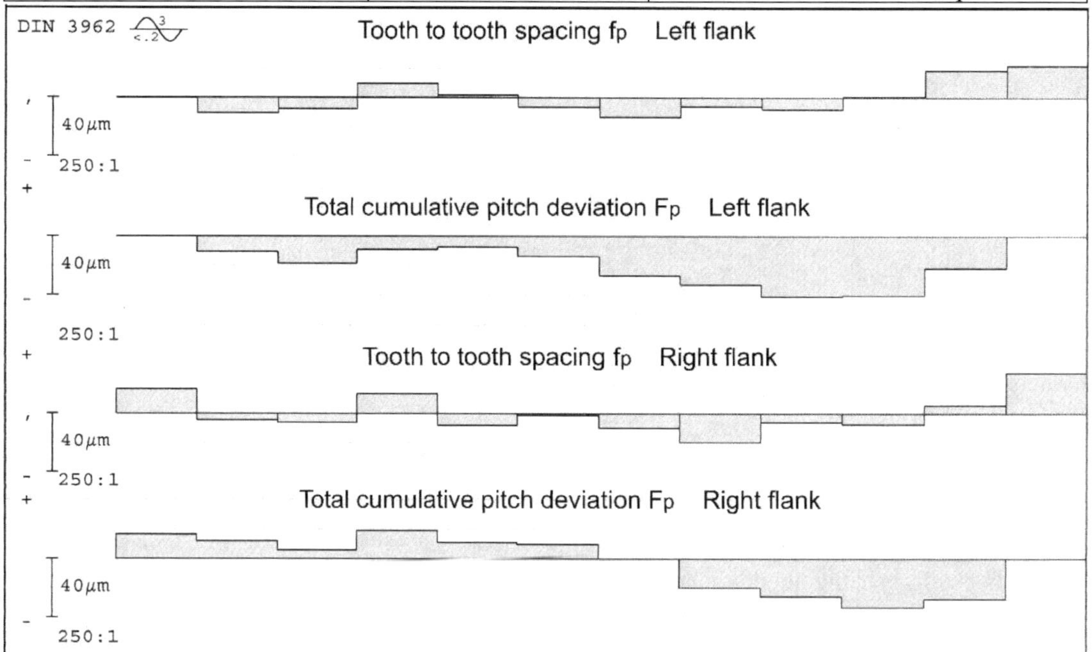

Pitch circle measure: 34.622 z=9mm	Left flank				Right flank			
	Measured value	Precision	Measured value	Precision	Measured value	Precision	Measured value	Precision
Max. single pitch deviation fp max	21.7	10	5.0	5	27.5	10	5.0	5
Max. difference between adjacent pitches fu max	21.9	9	6.0	5	22.0	9	6.0	5
Range of pitch error Rp	34.9				47.2			
Total cumulative pitch deviation Fp	40.9	9	16.0	5	52.1	9	16.0	5
Total cumulative pitch deviation z=8 teeth Fpz/8	40.2	10	10.0	5	44.1	10	10.0	5

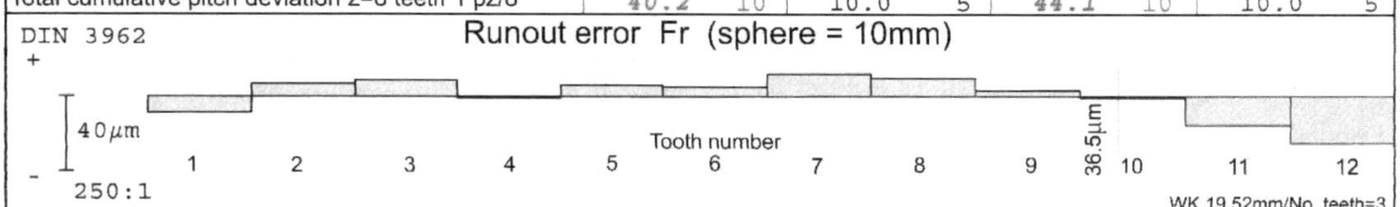

Runout error Fr	36.5	9	11.0	5	Limit value	19.47	19.43
Variation of tooth thickness Rs					Measured value	19.479	19.441

Figure 5. Individual and cumulative pitch deviations and runout

The measured values for F_r (figure 5), greater than those prescribed, generate noise during the pump's operation. The measured value of F_r = 36.5 µm corresponds [15] to the 9th grade of precision. The requirements for precision of this spur gear pump gearing is that it should correspond to the precision grade 5, at the imposed value of F_r = 11 µm.

According to the laboratory measurements using the Klingelnberg P150 centre for the driven gear z_2, there were found errors and similar values to those of the pinion z_1.

We consider that this gearing, from the point of view of its functional role, shall meet the criterion of smooth operation. The limit deviation tolerance for F_r is determined [14, 15] depending on the precision grade of the gearing.

The evaluation of the pump's gearing precision also required the evaluation of the specific deviations [16] regarding the profile and the flank line for each of the two gears (figures 6 and 7).

Name: Pinion	No. of teeth: 12	Pressure angle: 20°00'00"
Drawing no. Pinion z_1=12	Module m=2.54 mm	Flank line inclination angle: 0°0'0"
Order no. 01	Base circle diameter d_b=28.6418 mm	Face width b=18.3 mm Profile evaluation range L_a=9.28 mm
Condition: final		Flank line length L_b=14.64 mm
	Temperature: 21.4°	Probe rod ball diameter: 2 mm

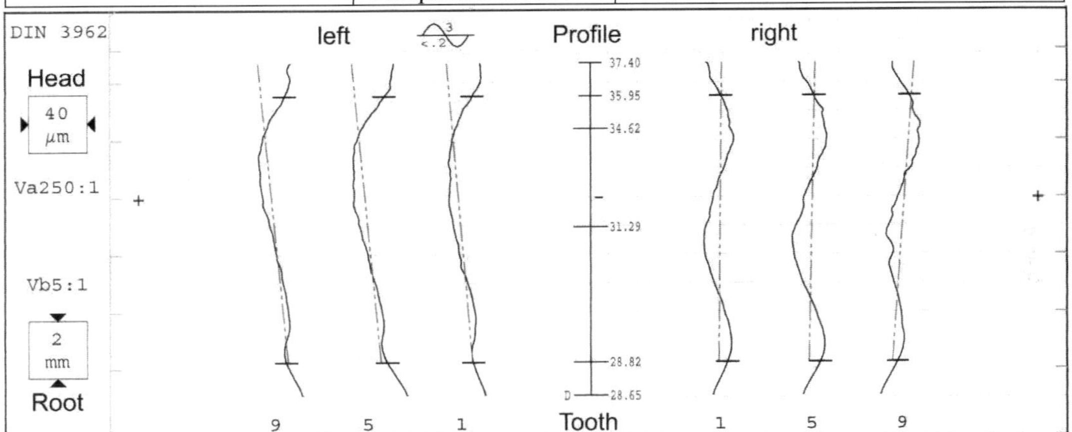

		Measured value [µm] Precision			Limit value Precision				Measured value [µm] Precision			
fHαm	17.2 9	V 2.6							6.6 7	V 12.1		
fHα		18.3 10	17.6	15.7	±4.5	5	±4.5	5	1.9	3.9	14.0 9	
Fα		21.3	23.2 9	18.8	7	5	7	5	20.5	23.8	24.6 9	
ffα		22.7	28.1 10	22.4	6	5	6	5	19.8	23.9 10	19.0	
P/C-φ [mm]									37.346			[37.3/37.4]

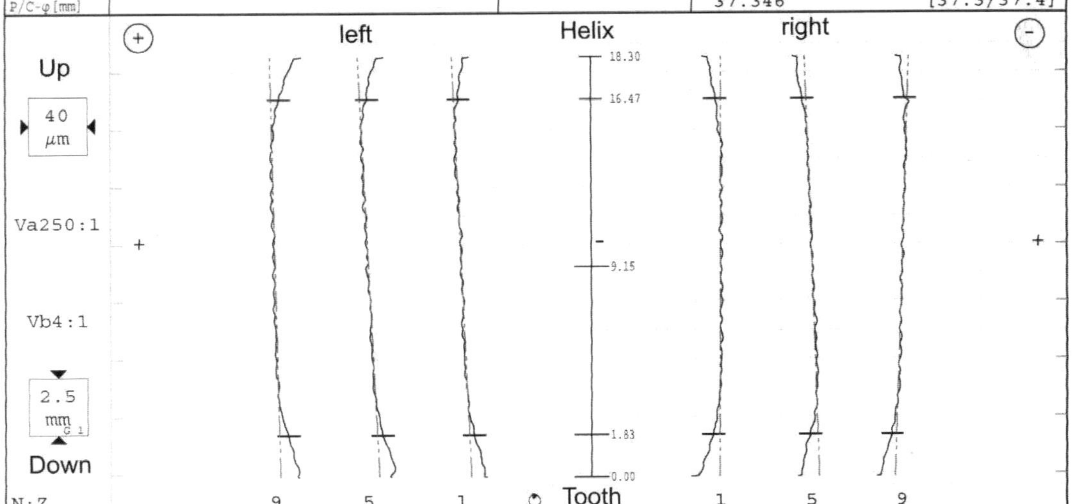

fHßm	11.6 8	V 7.6							-0.4 1	V 19.5		
fHß		6.9	14.5 8	13.3	±6	5	±6	5	-1.1	9.7	-9.8 7	
Fß		12.3	14.9 8	13.7	7	5	7	5	6.3	10.5	12.7 7	
ffß		8.7 8	7.2	5.5	4.5	5	4.5	5	6.1	6.8 7	5.2	

Figure 6. Diagrams and deviations of the teeth profile and of the flank line – pinion

Name: Gear	No. of teeth: 12	Pressure angle: 20°00'00"
Drawing no. Driven gear z_2=12	Module m=2.54 mm	Flank line inclination angle: 0^0
Order no. 01	Base circle diameter d_b=28.6418 mm	Face width b=18.3 mm Profile evaluation range L_α=9.28 mm
Condition: final		Flank line length L_b=14.64 mm
	Temperature: 21.3°	Probe rod ball diameter: 2 mm

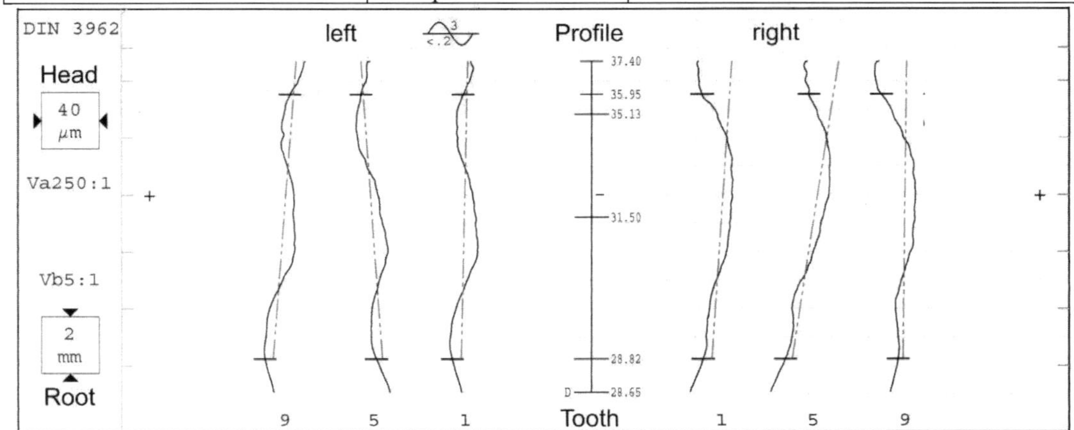

	Measured value [µm]	Precision			Limit value	Precision			Measured value [µm]	Precision		
fHαm	-1.9 3	V 27.6							14.4 9	V 26.6		
fHα		-14.5 9	13.1	-4.2	±4.5	5	±4.5	5	12.2	28.8 11	2.2	
Fα		21.5	22.0 9	19.7	7	5	7	5	21.2	30.9 10	23.7	
ffα		16.0	16.1	18.2 9	6	5	6	5	25.7 10	24.7	24.5	
P/C-φ [mm]									37.359		[37.3/37.4]	

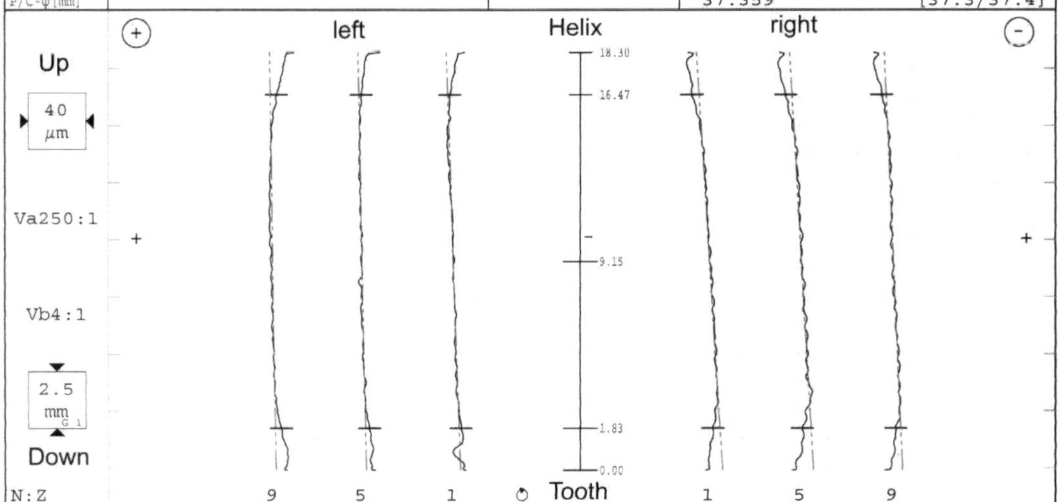

fHßm	6.9 6	V 5.7							16.1 9	V 5.2		
fHß		4.5	5.9	10.2 7	±6	5	±6	5	18.3 9	16.8	13.1	
Fß		8.7	7.7	10.0 7	7	5	7	5	18.1	18.9 9	13.4	
ffß		6.5 7	5.6	3.2	4.5	5	4.5	5	6.9	8.7 8	4.9	

Figure 7. Diagrams and deviations of the teeth profile and of the flank line – driven gear

The measured values refer to: $f_{H\alpha m}$ – mean profile slope deviation (arithmetic mean of the profile slope deviation of three or more equally spaced tooth flanks around the gear's circumference), $f_{H\alpha}$ – profile slope deviation/profile angular error (distance between two nominal profiles that intersect the average profile at start and end points of the profile range), F_α – total profile deviation/profile total error (distance between two nominal profiles enclosed within the profile test range, $f_{f\alpha}$ – profile form deviation/profile form error (distance between two involutes of the actual base circle, that enclose the actual involute profile within the profile inspection range), $f_{H\beta m}$ – mean helix slope deviation (arithmetic mean of the helix slope deviations of three or more equally spaced tooth flanks around the gear's circumference), $f_{H\beta}$ – helix slope deviation/tooth alignment tolerance (distance in transverse plane between two nominal leads that intersect the average lead – helix at start and end points of the lead inspection range), F_{β} – total helix deviation/lead total error (distance between the two nominal leads enclosed within the lead inspection range), $F_{f\beta}$ – helix form deviation/lead form error (distance between two helical lines that enclose the actual lead within the lead inspection range).

In figure 6, the total helix deviation of the tooth direction (flank line) F_{β} (for teeth 1, 5 and 9, left-hand measurement of the pinion z_1) has the values 13.7 µm, 14.9 µm and 12.3 µm, according to the criterion of contact between the teeth. These measured values correspond to precision grades 7 or 8 [14, 15], and the prescribed value (precision grade 5) for F_{β} deviation is 11 µm.

The measuring on each tooth profile is properly performed according to the tooth height, between the base circle and the head circle. The highest measured value has the deviation $f_{f\alpha}$ = 28.1 µm for the 5th tooth on the left flank (figure 6), which corresponds to the precision grade 10, a value that cannot be accepted because it generates an inappropriate operation of the gearing. Alongside are presented, for comparison, the limit values of deviations and tolerance F_{α} for the 5th grade of precision. Also, the measured values are indicated for the right flank. Figure 7 presents measurements for the teeth 1, 5 and 9 of the driven gear z_2. The figure also shows the highest value of the tolerance F_{α} = 30.9 µm, which corresponds to the precision grade 10. The analysis of the profile shapes for the two gears presents a certain similarity, corresponding to the left and to the right flank. We believe that this may be due to the generating profile of the tool's cutting edges, but also to the generating process of the teeth profile.

On diagrams from figures 6 and 7 are identified some specific points of the profiles measured on the left and right flank for those three teeth considered. Thus, it is possible to identify, without being noted, the following characteristic points [11]: the start of the measurement (diameter 28.65 mm), the start of the generated profile (diameter 28.82 mm), the start of flanking (diameter 34.6 mm - pinion and 35.13 mm – driven gear), of singular gearing on the tooth head (diameter 35.95 mm) and of profile limit on the tooth head (diameter 37.4 mm). From the profiles diagrams we observed that it was applied a modification by flanking on the left flanks to the pinion and on the right flanks to the driven gear. On the other corresponding flanks, right on the pinion and left on the gear it is also applied a modification called roundness on the tooth width.

According to [13, 15] for evaluation of the flanks direction deviations (figures 6 and 7) there were measured every three flanks left - right on pinion and on driven gear. For these deviations there are indicated actual values and limit values. Two areas are identified on every gear's face width (b = 18.3 mm): the first at the flanks ends (each on a 1.8 mm width) and the second on a length ($L_b \approx$ 15 mm). The active area of the flanks presents small deviations corresponding to the precision grades 7 and 8, which we consider that are acceptable. The flank on its width may be considered to have a slight longitudinal swell modification, which leads to a functional improvement. Also, this modification may diminish the deviations influence on axes not being parallel [17].

4. Conclusions

The measurement of the geometrical elements of the two gears and their deviations was performed with a specialized precision equipment, according to the methodology indicated in standards [14, 15]. The graphical and numerical results indicate that the two gears' teeth are corrected and correspond to the measured data and to the geometrical calculations made using the known relations [7, 11].

The measured deviations have some values that are different from those prescribed in the technical documentation [3]. The values of the precision grades of the limit tolerances and deviations have some common causes: constructive, of manufacturing and of assembly. To these are added the wear of the cutting tools, the inaccuracy of the gear cutting machine-tool [18], the positioning of the tool on the work surface, the speeds of the generation movements, the inaccuracy or lack of rigidity of the mechanical orientation, fixture and driving devices. The choice of the precision grades and of the teeth flanks roughness [19] should correspond to the pump's operating conditions. The deviations found, with values higher than those allowed for this type of pumps, determine the occurrence of inconsistencies in operation (vibration, noise, wear of the teeth flanks and of the bearing blocks). The beneficiary of a lot of such pumps noticed these problems and wanted to know their cause.

A next stage of research may consist in the elaboration of a measurement methodology of many other parameters with the pump in operation (the required driving power, noise and vibration levels, correlations between the viscosity of the hydraulic fluid and the driving torque, etc.). Consideration will also be given to the determination of the optimal variant of the technological processing parameters.

5. References

[1] Zhao X and Vacca A 2017 Formulation and optimization of involute spur gear in external gear pump *Mech. Mach. Theory* **117** 114-132
[2] Egbe E A 2013 Design analysis and testing of a gear pump *Int. J. of Eng. & Sci.* **3** 01–07
[3] UM Hidraulica Plopeni; Hesper 2019 Gear pumps & motors, *Company catalogs*
[4] Mucchi E, D'Elia G and Dalpiaz G 2012 Simulation of the running in process in external gear pumps and experimental verification *Meccanica* **47** 621–637
[5] Nagamura K, Ikejo K, Toyoshima T and Ueda M 2001 Design and performance of gear pumps with non-involute tooth profile *JSME Intern. Conf. on Motion & Power Transm.* 689-694
[6] Mucchi E, Dalpiaz G and Rivola A 2010 Elastodynamic analysis of a gear pump. Part II: Meshing phenomena and simulation results *Mech. Syst. & Signal Process.* **24(7)** 2180-2197
[7] Oprean A, Marin V and Dorin Al 1976 *Hydraulic power systems* (in Romanian: *Acţionări hidraulice*), (Editura Tehnică Bucureşti)
[8] Ghionea I 2010 Research on optimization by simulation of the industrial products design *PhD. Thesis*, University Politehnica of Bucharest
[9] Casoli P, Vacca A and Berta G L 2008 Optimization of relevant design parameters of external gears pumps *Proc. 7th JFPS Intern. Symp. on Fluid Power* Toyama 277-282
[10] Colborne J R 1987 *The geometry of involute gears* (Springer Verlag)
[11] SDP/SI 2014 *Elements of Metric Gear Technology*
[12] Project POSDRU/159/1.5/S/138963 2014 Parameterized design of elements from integrated product systems and technological manufacturing structures, Univ. Politehnica of Bucharest
[13] ISO 18653 2011 *Gears – evaluation of instruments for the measurement of individual gears*
[14] STAS 6273-81 1981 *Cylindrical gears. Tolerances*. Romanian standard
[15] DIN 3962 1978 *Tolerances for cylindrical gear teeth.* Part 1: *Tolerances for deviations of individual parameters*, Part 2: *Tolerances for tooth trace deviations*
[16] Lianfu H, Changfeng F, Wang J and Tang W 2013 Outlier detection and correction for the deviations of tooth profiles of gears, *Meas. Sci. Rev.* **13**, *J. of Inst. of Meas. Sci.*, Slovak Academy of Sciences, Walter de Gruyter GmbH, Berlin
[17] Velex P and Maatar M 1996 A mathematical model for analyzing the influence of shape deviations and mounting errors on gear dynamic behaviour *J. of Sound & Vibration* **191** 629–660
[18] Rădulescu Gh, Miloiu Gh, Gheorghiu N, Munteanu C, Vişa Fl, Ionescu N, Popovici Vl, Dobre G and Raşeev M 1986 *Practical guide for machine-tools design* (in Romanian: *Îndrumar de proiectare în construcţia de maşini*) **III**, (Editura Tehnică, Bucureşti)
[19] Akbarzadeh S and Khonsari M M 2008 Performance of spur gears considering surface roughness and shear thinning lubricant *ASME, J. of Trib.* **130(2)** 021503

Acknowledgements
The work has been funded by the Sectoral Operational Programme Human Resources Development 2007-2013 of the Ministry of European Funds through the Financial Agreement POSDRU/159/1.5/S/.138963. The authors wish to thank Neptun S.A. Câmpina for performing the measurements and for the assistance granted in interpreting the results.

IOP Conf. Series: Materials Science and Engineering **682** (2019) 012006 doi:10.1088/1757-899X/682/1/012006

Strategic outlook in industrial design assessment based on product category

A E Crăciun

Faculty of Industrial Engineering and Robotics, University Politehnica of Bucharest, 060042, Bucharest, Romania

craciun_alexandra@outlook.com

Abstract. Assessing industrial design is an intricate process of product analysis which aims to establish the quality degree of a product and considers all its implications from technical restrictions to appearance. Industrial design is related to mass produced articles obtained through industrial processes and whether products are articles of adornment or machines, both technical and aesthetic dimensions are relevant and influence the design process. All types of industrial products can be assessed using a numerical method based on the same criteria, however the same principles have a different impact on various types of products. The author proposes a customized approach in order to achieve an objective assessment for different concepts of design.

1. Introduction

The major dilemma in assessing the design of industrial products is given by the ambiguity of setting the general domain to which the design section belongs to. Being associated with industrial processes, design tends to become an integrant part of the technical aspects of production. At the same time, the term design is often used when referring to the appearance of a product considering the aesthetics or the significance of its shape or color.

Product design is strongly correlated with the industrial domain and thereby with mass production, a product is considered industrial due to the processes through which it is obtained, regardless of the parameters related to its typology, the exterior appearance or its components. A reference representation of the types of industrial products is the Locarno classification, where the products are divided into thirty-two classes and include both decorative products and industrial machines.

Regardless the nature of the product, it is meant to fulfill a well-defined purpose. There are design assessment methods meant to determine the suitability of the product attributes in correlation with the purpose given to the product [1]. Most methods are focused on aesthetic aspects, involving the analysis of visual elements [2], establishing the attributes involved in the product aspect [3], or simply studying the aesthetics of products form [4]. Also, the relationship formed by the product and its context of use is analyzed, the connection between them is examined [1], or there is the case where the areas of interest and the given context are established [5].

Based on all the aspects involved in the user experience, the quality of the product as well as the quality of the experience itself and the factors that determine the purchase decision, the author considers that the design assessment of industrial products needs to take into account principles regarding functionality, technical specifications, ergonomics, symbolistic and aesthetic aspects, referenced in this paper as design assessment dimensions.

IOP Conf. Series: Materials Science and Engineering **682** (2019) 012006 doi:10.1088/1757-899X/682/1/012006

2. Importance ratios of design dimensions

The author proposes a coherent design assessment method that implies understanding all the parameters involved in product design, based on exhaustive analysis of user experience that starts from safety and the implicit technical restrictions, the details that facilitate the use process as much as possible, symbolic implications in the product-user relation, as well as the aesthetics of the product interface. All the considered aspects in the assessment method are applicable on any product category, yet the differentiation between the type of products must also be reflected in their approach in design assessment.

Considering the design dimensions proposed by the author, Functional, Technical, Ergonomic, Symbolistic and Aesthetic, when assessing design for different product categories, the proposed method consists of a five steps structure of importance relevant to each type of category separately, whereas for a certain product category, a dimension such as Aesthetic or Ergonomic may be of much greater importance than for other category.

In order to determine the ratio values for each dimension, as seen in figure 1, a medium value of importance p_3 was established by dividing the total importance P to the number of design assessment dimensions n, as in equation (4), the value P being set to 100%, and represents the sum of all ratios, as observed in equation (1) or (2).

$$P = \sum_{n=1}^{5} p_n = 100\% \tag{1}$$

$$P = \frac{(p_1 + p_n) * n}{2} \tag{2}$$

$$p_n = \frac{p_{n-1} + p_{n+1}}{2} \tag{3}$$

$$p_3 = \frac{P}{n} = \frac{100\%}{5} = 20\% \tag{4}$$

P, the total importance value, n is the total number of assessment dimensions and p_3 represents the medium ratio.

The mean value of the series is considered the average value between the minimum and maximum ratio, in this case p_3. In order to keep the ratio constant, a p_n value represents the average between the two adjacent ones as observed in equation (3).

Considering the medium value of importance of a ratio p_3 and the relations between the terms of the sum, as in equation (3), all other values for the ratios were determined and presented in figure 1.

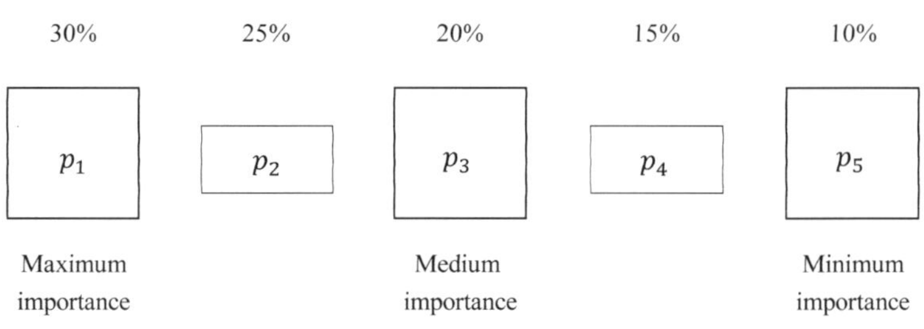

Figure 1. Factors of importance for the five assessment dimensions.

It is relevant to have an importance difference between the dimensions but the gap should not be too extensive as all the proposed dimensions are relevant in the actual assessment. As seen, the maximum importance of a dimension is established at 30% and the minimum at 10%, for all these ratios, in order to carry out the assessment it is necessary to establish the correlation with the proposed dimensions.

3. Dimensions hierarchy for the Locarno classes

Starting from the proposed assessment dimensions, using the Locarno classification, the correspondent structure according to the importance degree of the dimension is established for each class. All the proposed dimensions have an impact on the assessment of an industrial product, however the extent to which they impact the product differs.

3.1. The structure of design assessment dimensions

The functional assessment involves the actual testing of the product and establishes the degree to which the product is adequate to the use experience. Therefore, functionality is the most important aspect in product design, it is essential to be considered a primary dimension regardless of the product class.

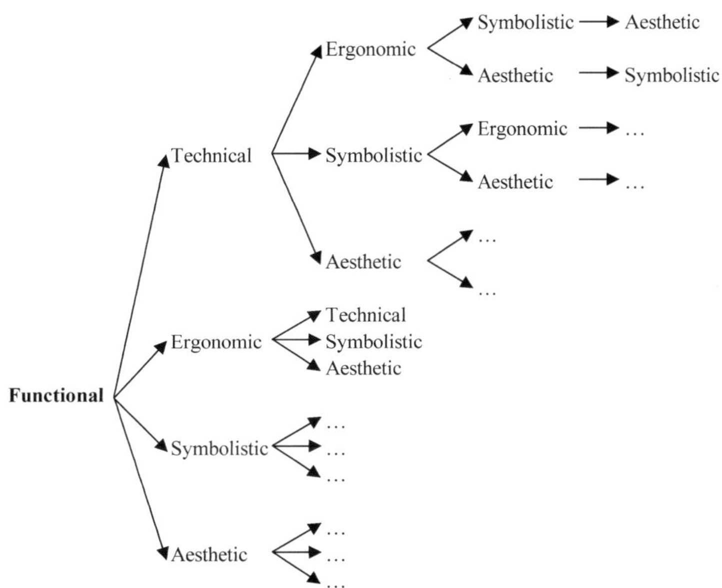

Figure 2. The hierarchical structure of the assessment dimensions, proposed by the author.

When calculating the number of variations, D_1 is represented by the functional dimension, its ratio p_1 being a constant (30%) and not a variable as is the case for the other dimensions. Established according to importance, the position of only the four dimensions may vary, as presented in figure 2. Based on Functional as primary design dimension, there are twenty-four possibilities for ranking the assessment dimensions as seen in equation (5).

$$C_4^3 + C_3^2 + C_2^1 = 24 \quad (5)$$

3.2. Corresponding classes for each importance layout

Considering the nature of industrial products, these being articles designed with a well-defined purpose, the aspects of functionality are of major importance, any other dimension involved in the design assessment being secondary.

In the proposed method, the hierarchy for a particular class of products involves analyzing the product typology and establishing the aspects with the greatest impact on the design for the given category. The thirty-two classes were divided into four categories for the secondary dimensions, in three remaining categories for the tertiary dimensions and so on until the remaining fifth dimension and for each variation of importance the corresponding classes were established.

IOP Conf. Series: Materials Science and Engineering **682** (2019) 012006 doi:10.1088/1757-899X/682/1/012006

3.2.1 Technical as a secondary dimension
As presented in table 1, the industrial products that are defined by technical as secondary dimension are mainly products used in the industrial field such as Tools and hardware (Class 8), Means of transport or hoisting (Class 12) or Machines (Class 15). The classes of products corresponding to this category are easily identifiable because the technical aspects are all the more obvious as the product is related to the technology and the designer must have engineering knowledge, therefore, aspects related to sizing tolerances, material properties or standard colour coding must be considered as they have a direct impact on safety in use.

Table 1. Hierarchy of classes with secondary Technical dimension

Third dimension	Fourth dimension	Fifth dimension	Corresponding Locarno class
Ergonomic	Symbolistic	Aesthetic	8, 12, 13, 22, 29
	Aesthetic	Symbolistic	15, 16, 23, 24, 31
Symbolistic	Ergonomic	Aesthetic	-
	Aesthetic	Ergonomic	-
Aesthetic	Ergonomic	Symbolistic	-
	Symbolistic	Ergonomic	-

3.2.2 Ergonomic as a secondary dimension
Ergonomic dimension is strongly correlated with the actual process of use of a product and facilitating that certain experience as much as possible. Products defined by great importance in ergonomics primarily involve a proper sizing compared to the dimensions of the human body and take into account the entire experience of use from the effective handling of the product to the direct impact on the user's well-being during use experience. As seen in table 2, among the classes of products corresponding to this category are listed Household goods (Class 7) or Packages and containers for the transport or handling of goods (Class 9).

Table 2. Hierarchy of classes with secondary Ergonomic dimension

Third dimension	Fourth dimension	Fifth dimension	Corresponding Locarno class
Technical	Symbolistic	Aesthetic	9, 18
	Aesthetic	Symbolistic	4, 14, 21, 25, 28, 30
Symbolistic	Technical	Aesthetic	-
	Aesthetic	Technical	3
Aesthetic	Technical	Symbolistic	6, 20
	Symbolistic	Technical	7,19

3.2.3 Symbolistic as a secondary dimension
Symbolism prevails over other proposed dimensions if the products are defined by historical elements with a strong symbolism already established. The specific products can determine the creation of other elements of artistic nature or they could be intended for hobbies. This is the narrowest category mainly because this type of products is usually obtained by handicraft and not industrial design as is the case with Musical instruments (Class 17), presented in table 3.

IOP Conf. Series: Materials Science and Engineering **682** (2019) 012006 doi:10.1088/1757-899X/682/1/012006

Table 3. Hierarchy of classes with secondary Symbolistic dimension

Third dimension	Fourth dimension	Fifth dimension	Corresponding Locarno class
Technical	Ergonomic	Aesthetic	17
	Aesthetic	Ergonomic	-
Ergonomic	Technical	Aesthetic	-
	Aesthetic	Technical	-
Aesthetic	Technical	Ergonomic	32
	Ergonomic	Technical	27

3.2.4 *Aesthetic as a secondary dimension*

Although they are produced trough industrial processes, for certain articles, aesthetics is defining and has the greatest impact on the purchase decision. Aesthetic as a secondary dimension is specific for products such as Articles of clothing and haberdashery (Class 2) or Textile piece goods, artificial and natural sheet material (Class 5), as presented in table 4.

Table 4. Hierarchy of classes with secondary Aesthetic dimension

Third dimension	Fourth dimension	Fifth dimension	Corresponding Locarno class
Technical	Ergonomic	Symbolistic	10
	Symbolistic	Ergonomic	26
Ergonomic	Technical	Symbolistic	1
	Symbolistic	Technical	-
Symbolistic	Technical	Ergonomic	-
	Ergonomic	Technical	2, 5, 11

4. Hierarchical structure of the dimensions in correlation with the corresponding ratios of importance

In order to carry out the assessment, as in equation (6), the proposed ratios p_n must be correlated with the corresponding dimensions D_n, and, for each dimension, its hierarchy is established within the assessment according to the product category.

$$T_D = \sum_{n=1}^{5} D_n p_n = D_1 p_1 + D_2 p_2 + D_3 p_3 + D_4 p_4 + D_5 p_5 \quad (6)$$

T_D, the total value of design assessment, n is the total number of assessment dimensions, D_n represents the assessed dimension with its corresponding ratio p_n.

The total value of the assessment represents the sum of all the dimensions D_n related to the corresponding ratio p_n. The D_n values are given in the actual assessment of the product and if all the qualitative requirements of the product are met and the dimensions obtain maximum value, T_D can take values up to 100, which represents the total agreement in meeting the design requirements.

IOP Conf. Series: Materials Science and Engineering **682** (2019) 012006 doi:10.1088/1757-899X/682/1/012006

Table 5. Dimensions hierarchy and corresponding ratios for the ten most common Locarno classes

Class	Class description	Functional	Technical	Ergonomic	Symbolistic	Aesthetic
Class 2	Articles of clothing and haberdashery	30%	10%	15%	20%	25%
Class 6	Furnishing	30%	15%	25%	10%	20%
Class 7	Household goods	30%	10%	25%	15%	20%
Class 9	Packages and containers for the transport or handling of goods	30%	20%	25%	15%	10%
Class 11	Articles of adornment	30%	10%	15%	20%	25%
Class 13	Equipment for production, distribution or transformation of electricity	30%	25%	20%	15%	10%
Class 15	Machines	30%	25%	20%	10%	15%
Class 21	Games, toys, tents and sports goods	30%	20%	25%	10%	15%
Class 26	Lighting apparatus	30%	20%	10%	15%	25%
Class 30	Articles for the care and handling of animals	30%	20%	25%	10%	15%

The proposed method applies for all the subclasses corresponding to a certain class, even if the products in the subclasses have differentiations in the product typology, the ratios remain valid, Locarno being a classification of industrial products according to the destination of use.

5. Conclusions

The author has developed a new approach in industrial design assessment starting from establishing the importance ratios for the five dimensions and their hierarchy according to the product typology, which resulted in a strategic method of industrial design assessment that addresses all categories of industrial products depending on the impact of a dimension on the overall design.

When differentiating the ratios of dimensions for certain classes, a specific pattern can be observed for certain dimensions with high importance, for example there are far fewer classes with a very high degree of importance in symbolism and also the classes defined by high ratios for technical aspects are strongly correlated with ergonomics and less with aesthetics or symbolism. Also, if the product is assessed in the development stage, as observed, the product features correlated with the high ratio dimension prevail over those with low importance, for example, when a product from the Machines class is assessed in the concept phase, as seen in table 5, it is recommended that the features related to functionality and technical aspects must be fulfilled first, then the ergonomic details are established, followed by the aesthetics and only at the end improvements regarding the symbolism can be made. The proposed approach thus aims to ensure the high quality of any industrial product both by addressing all its key aspects as well as considering them in a systematic way that takes into account the specifics of the product according to its typology and the destination of use.

6. References

[1] Dumitrescu A 2013 *Tratat de Design* (Bucharest: Editura Academiei Române) p 445, p 609
[2] Warell A 2003 *Towards a theory-based method for assessment of visual form syntactics* (Göteborg: Chalmers University of Technology) p 4
[3] Blijlevens J, Creusen M E and Schoormans J P 2009 *Int. J. of Des.* **3** pp 27-35
[4] Birkhoff G D 1933 *Aesthetic Measure* (Massachusetts Harvard University Press) p 4
[5] Ulrich K T and Eppinger S. D 2014 *Diseño y desarrollo de productos* (McGRAW-HILL, 5th edition) p 223

IOP Conf. Series: Materials Science and Engineering **682** (2019) 012007 doi:10.1088/1757-899X/682/1/012007

The vibration behaviour of a freight wagon in the presence of irregularities of the track

T Mazilu[1] and V M Dinu[2]

[1] Department of Railway Vehicles, University Politehnica of Bucharest, 313 Splaiul Independenţei, Bucharest, Romania
[2] Doctoral School of Transports, University Politehnica of Bucharest, 313 Splaiul Independenţei, Bucharest, Romania

trmazilu@yahoo.com

Abstract. The paper presents an analysis of the behaviour of the vertical vibration of a freight wagon featured with dry friction dampers in suspension. To this end, the wagon model is reduced to an oscillator with two degrees of freedom corresponding to the suspended mass of the wagon and the wheel. The suspension has progressive characteristic in two steps according to the empty/loaded state. Coulomb's model is considered to model the dry friction damper. The vibration is induced by the track irregularity which is synthetized from an analytic spectrum. Vibration behaviour exhibits two different aspects: at low speed, the suspension is locked, and at high speed, the stick-slip vibration occurs. Acceleration of the suspended mass of the wagon and wheel-rail contact force increase along the speed. Wagon loaded state influences the wheel-rail contact force, but it has no impact upon the acceleration.

1. Introduction

Generally, during running, railway vehicles are subjected to a complex vibration behaviour that develops mainly in the vertical and lateral direction. The sources of these vibrations are very varied, but the main cause is the unevenness of the track. In the case of freight wagons, the vibrations affect the ride quality of the rolling stock, the fatigue resistance of the running gear and the track superstructure, jeopardizing, under certain conditions, the safety of traffic. At the same time, the dynamic loads that appear at the wheel-rail interface due to vibrations can lead to a reduction in the operating time of the track, its exit from the geometric parameters, etc.

From scientific point of view, freight wagons are modelled using non-linear oscillators due to the use of damping based on dry friction. This aspect raises interesting problems from the perspective of modelling the dry friction, related to the simulation of the possibility of the suspension blocking [1-2]. At the same time, the vibration behaviour can exhibit the stick-slip phenomenon, which sometimes has the characteristics of a chaotic motion [3].

In this paper, the vibration behaviour of a freight wagon due to the track irregularity is analysed using a simple model consisting of two-degree of freedom oscillator, representing the wheel and the suspended mass of the wagon. The wagon suspension is modelled with the help of an ideal spring working in parallel with a damper with dry friction. Such simple model has the advantage to allow making the analysis and observations easier.

2. Freight wagon model

Figure 1 shows the Y25 bogie of a Zas tank car. As it is known [4], the Y25 bogie suspension consists of two sets of double springs for each axle box with progressive elastic characteristic. Thus, up to 136 kN axle load, the elasticity of the bogie suspension is 0.3 mm/kN, and from this value, the elasticity becomes 0.1 mm/kN. The total mass of the bogie is about 4400 kg, of which the wheelsets mass is 2600 kg. The total mass of the wagon is 24000 kg for empty state and 80000 kg for loaded state. Each axle box is equipped with a Lenoir damping device.

Figure 1. Y25 bogie.

Figure 2 presents the proposed model for the study of the vertical vibrations of the tank car. The model is based on the symmetry property of the wagon construction, neglecting the coupling between the two bogies, as well as the coupling between the axles of the same bogie. At the same time, it is observed that the running track has a symmetrical construction with respect to the longitudinal axis. As a result, the wagon model consists of an oscillator with two degrees of freedom in which the mass M_c represents the suspended mass of the wagon that rests on one wheel (1/2 wheelset), and M_w - the mass of the wheel.

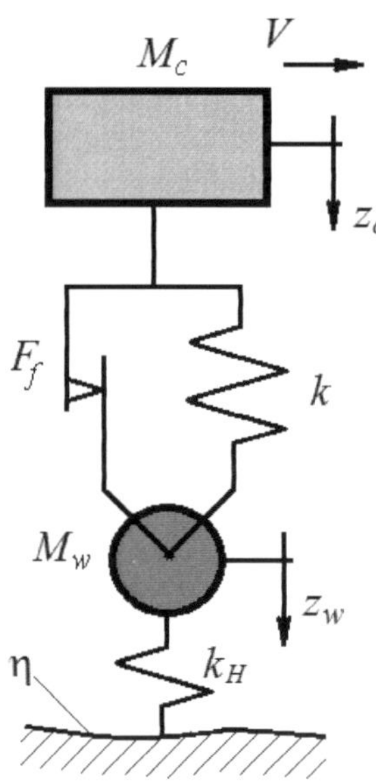

Figure 2. Mechanical model.

The stiffness of the suspension is k, and the frictional force developed by the dry friction damper is F_f. The track is assumed to be rigid with random irregularity η. It is assumed that the wheel of the vehicle is perfectly round, without geometrical deviations. The contact between the wheel and the rail is elastic, according to Hertz's non-linear theory of elastic contact between two bodies. The Hertzian contact is characterized by the Hertz C_H constant which depends on the main curvatures of the bodies in contact (wheel and rail) and the stiffness k_H. The oscillator moves with constant speed V. A similar model, but which considers the wheel-rail contact as rigid and harmonic track irregularity is presented in refs. [5-6].

Considering that the displacement of the suspended mass is z_c, and the displacement of the wheel is z_w, one can write the equations of motion of the model. Therefore, by applying the second principle of dynamics, we have:

- the equation of motion of the suspended mass

$$M_c \ddot{z}_c = F_e + F_f \tag{1}$$

- the wheel movement equation

$$M_w \ddot{z}_w = Q_o - Q - F_e - F_f , \tag{2}$$

wherein F_e is the elastic force in suspension

$$F_e = -k(z_c - z_w) , \tag{3}$$

F_f is the dry friction force, Q_o is the static load

$$Q_o = (M_c + M_w)g , \tag{4}$$

where g is the gravitational acceleration and Q is the wheel-rail contact force.

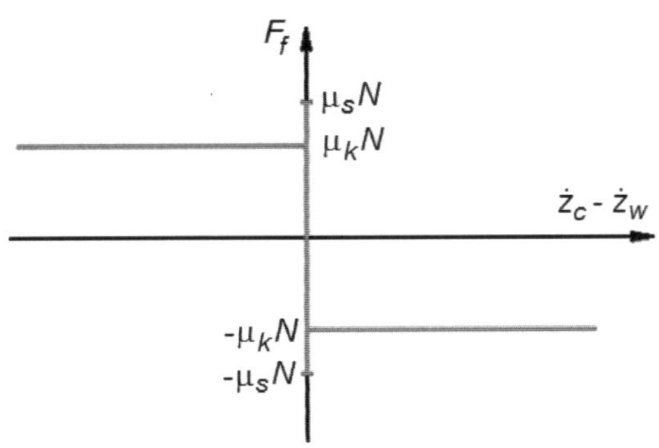

Figure 3. Friction force.

The friction force depends on the relative speed between the suspended mass and wheel and cannot exceed the static friction force (figure 3)

$$F_f = -\mu_k N \cdot \mathrm{sign}(\dot{z}_c - \dot{z}_w) \text{ for } \dot{z}_c - \dot{z}_w \neq 0 \tag{5a}$$

$$|F_f| \leq \mu_k N \text{ for } \dot{z}_c - \dot{z}_w = 0 \tag{5b}$$

where N is the normal pressing force in the damping element, μ_s is the static friction coefficient and μ_k - the kinematic friction coefficient.

The normal pressing force can be calculated with the relation

$$N = M_c \cdot g \cdot \mathrm{tg}\alpha , \tag{6}$$

where α is the inclination angle of the Lenoir mechanism; $M_c g$ represents the static force in the suspension spring and the variation due to the dynamic deflection of the spring is neglected.

The wheel-rail contact force can be calculated based on Hertz's theory

$$Q = C_H \sqrt{(z_w - \eta)^3} \cdot \sigma(z_w - \eta) , \tag{7}$$

where $\sigma(.)$ is the unitary step function of Heaviside.

The use of the unitary step function allows the modelling of the contact loss between the wheel and the rail: if z_w-η is positive, the contact force is positive, but this becomes zero when the condition z_w-$\eta \leq 0$ is met - the wheel has broken off the rail (figure 4).

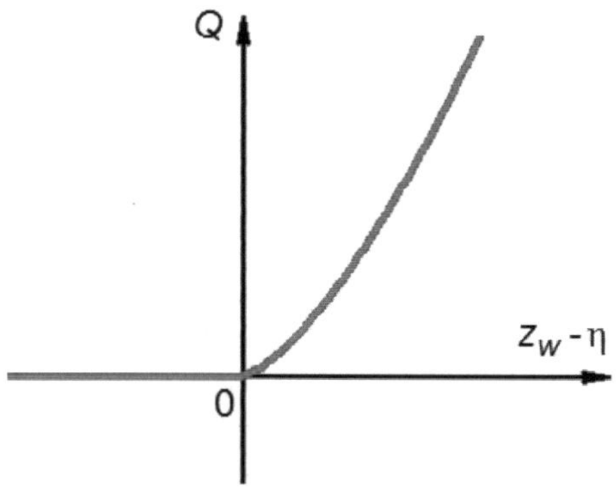

Figure 4. Wheel-rail contact force.

Suspended mass-wheel displacement is introduced

$$u = z_c - z_w , \tag{8}$$

and Eq. (1) becomes

$$M_c \ddot{u} = F_e + F_f - M_c \ddot{z}_w . \tag{9}$$

Wheel acceleration results from Eq. (2)

$$\ddot{z}_w = \frac{1}{M_w}\left(Q_o - Q - F_e - F_f\right). \tag{10}$$

Inserting wheel acceleration in Eq. (9), it results

$$\ddot{u} = \frac{1}{M_{ech}}\left(F_t + F_f\right), \tag{11}$$

where

$$M_{ech} = \frac{M_c M_w}{M_c + M_w} \tag{12}$$

and

$$F_t = F_e + \frac{M_{ech}}{M_w}\left(Q - Q_o\right). \tag{13}$$

IOP Conf. Series: Materials Science and Engineering **682** (2019) 012007 doi:10.1088/1757-899X/682/1/012007

Equations of motion (10-11) can be solved using a similar numeric procedure as it is presented in reference [6].

It should be noted that, since the wagon model is an oscillator with two degrees of freedom corresponding to the vertical motion of the suspended mass and the wheel, with two elastic levels, one for suspension, and the other for the elasticity of the wheel-rail contact, the vibration behaviour is characterized by two natural frequencies that are calculated with the relation

$$\nu_{1,2} = \frac{1}{2\pi}\sqrt{\frac{1}{2}\left[\frac{k}{M_c} + \frac{k+k_c}{M_w} \mp \sqrt{\left(\frac{k}{M_c} + \frac{k+k_c}{M_w}\right)^2 - 4\frac{kk_c}{M_c M_w}}\right]}. \tag{14}$$

On the other hand, the presence of dry friction can lead to the suspension locking, in which case the wheel moves in solidarity with the suspended mass, and the model 'degenerates' into one with a single degree of freedom whose natural frequency is given by the relation

$$\nu = \frac{1}{2\pi}\sqrt{\frac{k_c}{M_w + M_c}}. \tag{15}$$

Regarding the track irregularity, described by the function $\eta(x)$, where x is coordinated along the track, this can be determined either experimentally or by calculation based on the spectra obtained by measurements or their analytical approximations.

The longitudinal level can be calculated starting from the power spectral density of the longitudinal level, recommended by the ORE report B176 [7] as representative for the European rail networks,

$$S(\Omega) = \frac{A\Omega_c^2}{(\Omega^2 + \Omega_r^2)(\Omega^2 + \Omega_c^2)}, \tag{16}$$

where Ω is the wave number, Ω_c = 0.8246 rad/m, Ω_r = 0.0206 rad/m, and A is a coefficient depending on the quality of the tread, and applying the method recommended in ref. [8].

3. Numerical applications

In this section are presented a series of results obtained by applying the model and method presented above in the MATLAB environment. The parameters of the vehicle model are [4]: M_c = 2350 kg for empty wagon and M_c = 9350 kg for loaded wagon, M_w = 650kg, k = 830 kN/m for axle loads up to 135 kN and k = 2.5 MN/m for axle loads greater than 135 kN, μ_k = 0.17, μ_s = 0.204, N = 9.126 kN for the empty wagon and N = 36.699 kN for the loaded wagon, the inclination angle of the Lenoir mechanism,α = 21.42°, C_H = 118.6 GNm$^{3/2}$ and Δt = 1/20000 s.

The wheel load is 30 kN for the empty wagon and 100 kN for the loaded wagon. The wheel-rail contact rigidity is 1.125 GN/m for the empty wagon and 1.681 GN/m for the loaded wagon. The static friction force is 1.882 kN for the empty wagon and 7.486 kN for the loaded wagon, and the kinematic friction force is 1.568 kN for the empty wagon and 6.239 kN for the loaded wagon.

Figure 5 shows the longitudinal level of the track calculated for 1000 m section, including wavelengths between 3 m and 120 m. The peak value reaches 7.4 mm, while the RMS value is 2.38 mm which encloses the section of track in QN2 quality level for speeds between 80 and 120 km/h.

Figure 6 shows the friction force and the elastic force in the loaded wagon suspension at 20 km/h. It is observed that the friction force does not exceed 500 N, a value much lower than the static friction force (7.486 kN). This means that the suspension is locked, and this aspect is also signalled by the fact that the elastic force is zero.

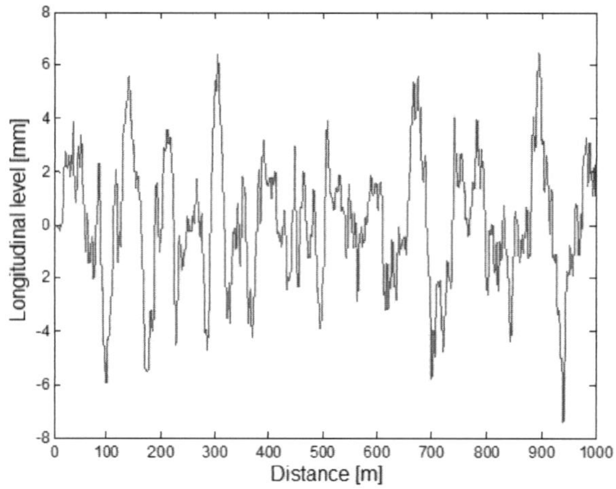

Figure 5. Longitudinal level.

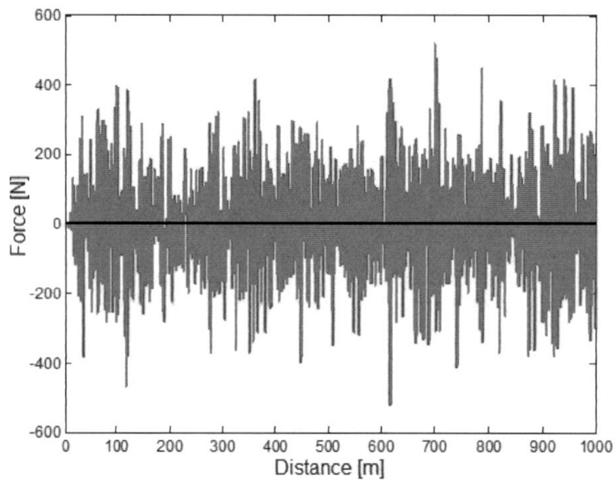

Figure 6. Friction force and elastic force in the loaded wagon suspension at 20 km/h: red line – friction force, black line – elastic force.

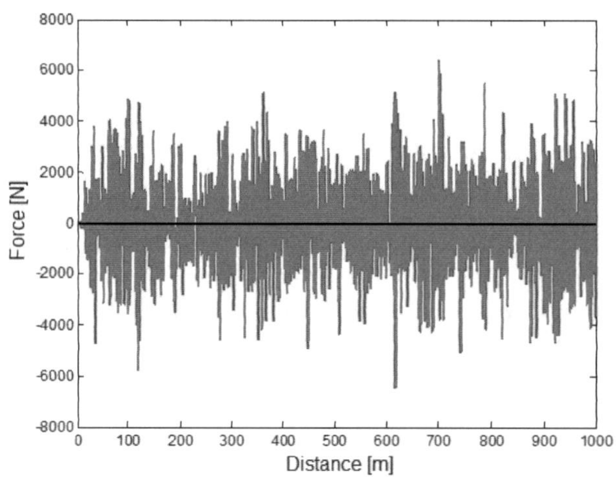

Figure 7. Friction force and elastic force in the loaded wagon suspension at 70 km/h: red line – friction force, black line – elastic force.

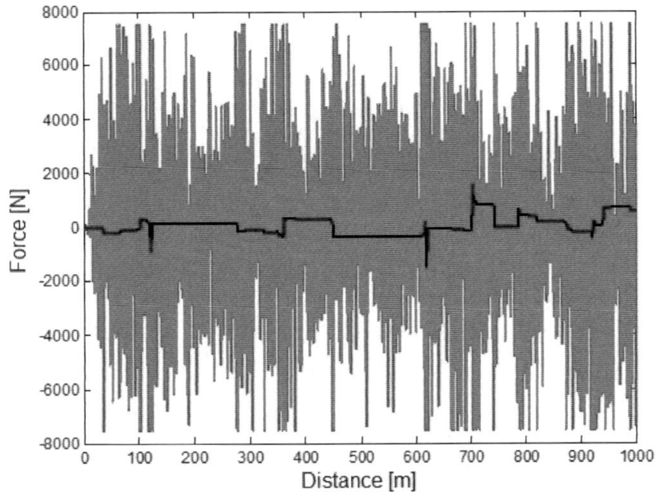

Figure 8. Friction force and elastic force in the loaded wagon suspension at 90 km/h: red line – friction force, black line – elastic force.

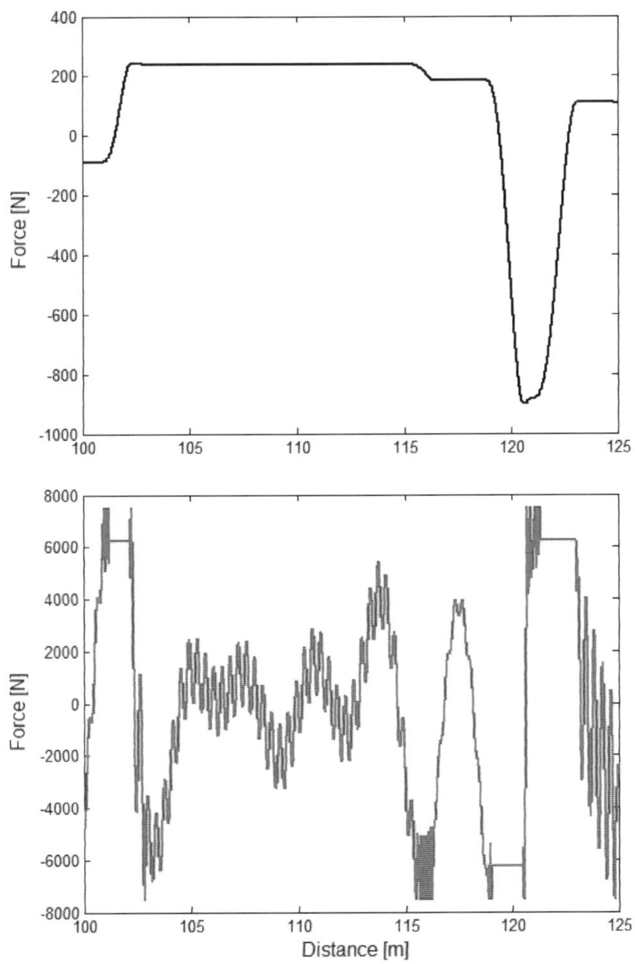

Figure 9. Stick-slip vibration in the loaded wagon suspension at 90 km/h: red line – friction force, black line – elastic force.

Similar diagrams also result in the simulation of the wagon circulation with speeds up to 70 km/h, as shown in figure 7, where the calculation speed is even 70 km/h. The difference is that the friction force is much higher than in the previous example; the peak value reaches 6.414 kN, but this is still lower than the static friction force, which explains why the suspension is locked throughout the simulation.

Figure 8 shows the same forces for the loaded wagon at 90 km/h. This time, the friction force often reaches the static friction force, and the suspension begins to work. This is consistent with the fact that the elastic force is no longer zero. It is interesting what happens when the friction force reaches the static friction threshold and the sliding between the wheel and the suspended mass occurs (see fig. 9). We observe the alternation of stick and slip phases - stick-slip vibrations. During the stick phase, the suspension is locked, the elastic force is constant, and the friction force has a reduced frequency variation, corresponding to the wagon's natural frequency with the locked suspension (approx. 65 Hz) modulated with the frequency induced by the longitudinal level of the track.

During the slip phase, the elastic force varies from one constant value to another, while the friction force has a high frequency variation when it goes from the static friction value to the dynamic friction value. If the slip phase is long enough, then the frictional force stabilizes to the value of the kinematic friction force.

From the point of view of the ride quality, the acceleration of the suspended mass (figure 10) and the wheel-rail contact force (figure 11) are of interest. The peak value of the suspended mass acceleration is 0.931 m/s^2, and its RMS value is 0.327 m/s^2. The wheel-rail contact force has variations of not more than 10.02 kN around the static load on the wheel (100 kN for the loaded wagon), while the RMS value is 3.28 kN.

The ride quality of the vehicle depends on the state of loading and the running speed. Figures 12 and 13 show the RMS value of the acceleration and the contact force respectively for the speed between 20 and 120 km/h for both empty and loaded wagon. It is observed that in the case of acceleration of the suspended mass, we do not notice differences depending on the condition of the wagon, empty - loaded. RMS acceleration increases with the running speed, reaching 0.5 m/s^2 at 120 km/h. The shape of the acceleration curve changes when the suspension starts working, around 70 km/h. The contact force increases with the speed of the wagon and is significantly higher in the case of the loaded wagon. For example, at 120 km/h, the RMS value of the contact force is 5.23 kN for the loaded wagon and only 1.58 kN if the wagon is empty. There is also a change in the allure of the contact force curve when the suspension starts working.

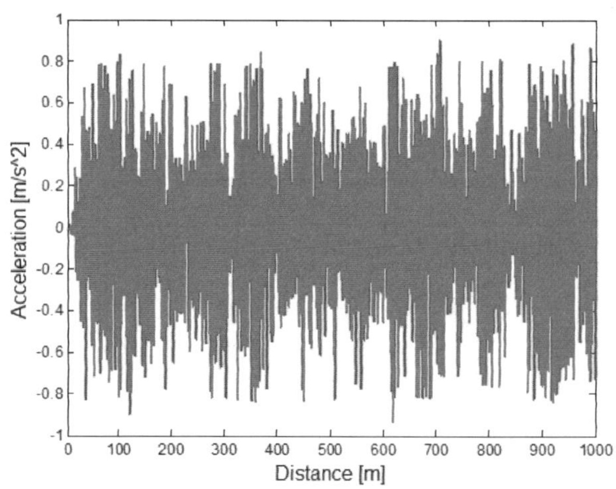

Figure 10. Acceleration of the suspended mass of the loaded wagon at 90 km/h.

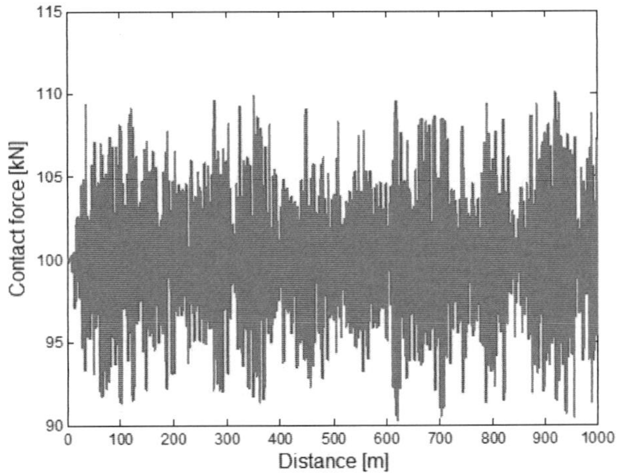

Figure 11. Wheel-rail contact force at 90 km/h (loaded wagon).

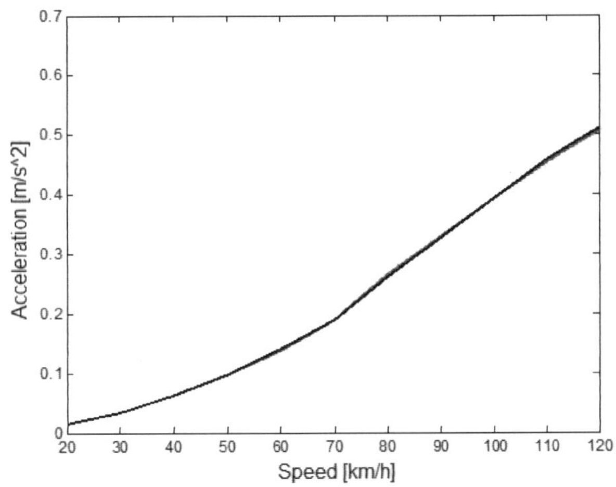

Figure 12. Acceleration of the suspended mass of the wagon:
red line – empty wagon; black line – loaded wagon.

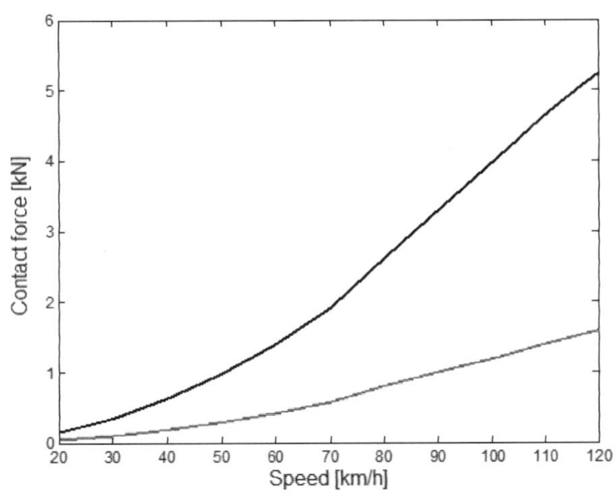

Figure 13. Wheel-rail contact force:
red line – empty wagon; black line – loaded wagon.

4. Conclusions

The paper presents a series of characteristics of the vertical vibrations of a freight wagon equipped with suspension with dry friction dampers when circulating on a track in the presence of longitudinal level deviations. For this, the wagon was modelled with the help of an oscillator with two degrees of freedom representing the suspended mass of the wagon and the wheel. The suspension is progressive in two levels for the empty and respectively charged state. For dry friction, Coulomb's model was considered.

At low speeds up to 70 km/h, the suspension of the wagon is locked, so that, at higher speeds, relative displacements occur between the suspended mass and the wheel in the form of stick-slip vibrations. The stick-slip vibrations have two distinct phases: the stick phase and the slip phase. During the stick phase, the friction force shows variations in the resonance frequency of the vehicle on the elasticity of the wheel-rail contact that are modulated by the longitudinal level deviations; the elastic force is constant, and the suspension does not work. During the slip phase, with the release of the suspension, high frequency vibrations may occur.

From a practical point of view, the acceleration and the wheel-rail contact force are of interest. These quantities increase continuously with the wagon speed after a different allure at the speeds at which the suspension is no longer locked. The results of the numerical simulations showed that the loading state does not influence the RMS acceleration level of the suspended mass. Consequently, the ride quality is not much influenced by the loading which is an advantage offered by the wagon suspension. This is explained by the fact that the wagon's natural frequencies do not differ much from the loading state due to the stiffness of the suspension. The contact force is influenced by the loading condition, being higher in the case of the loaded wagon than in the empty wagon in a ratio approximately equal to that of the static load on the loaded/empty wheel (in the analysed case 3.33: 1).

5. References

[1] Karnopp D 1985 Computer simulation of stick-slip friction in mechanical dynamic systems *J. Dyn. Syst-T ASME* **107** 100–103
[2] Oprea R A, Tudorache C M and Stan C 2015 A switch model for the study of powered axles stick-slip *Int. J. Heavy Veh. Syst.* **22** 338-355
[3] Popp K and Stelter P 1990 Stick-slip vibrations and chaos *Philos. Trans. R. Soc. London A* **332** 89-105.
[4] Burada C, Buga M and Crăsneanu A 1980 *Elements and structures of railway vehicles*, in Romanian (Bucureşti: Ed. Tehnică)
[5] Mazilu T and Dinu V M 2017 On the stick-slip vibration in the suspension of a freight wagon *MATEC Web Conf.* **112** 07013
[6] Dinu V M 2018 Numerical integration of the equation of motion of an oscillator with dry friction *IOP Conf. Ser. Mater. Sci. Eng.* **444** 042006
[7] ORE B 176:1989 Bogies with steered or steering wheelsets *Report No. 1: Specifications and preliminary studies*, vol. 2. Specification for a bogie with improved curving characteristics
[8] Dumitriu M 2014 Method to synthesize the trac vertical irregularities *Scientific Bulletin of the „Petru Maior" University of Tîrgu Mureş* **11(2)** 17-24

IOP Conf. Series: Materials Science and Engineering **682** (2019) 012008 doi:10.1088/1757-899X/682/1/012008

Theoretical calculus of the mechanical strength of binary oxide glasses

M Eftimie[1], A Filip[2] and A Banu[3]

[1]University Politehnica of Bucharest, Faculty of Applied Chemistry and Materials Science, 1 Polizu Str., Bucharest, Romania
[2]National Institute for Laser, Plasma and Radiation Physics, 409 Atomistilor Str., Magurele, Romania
[3]University Politehnica of Bucharest, Manufacturing Engineering Department, 313 Splaiul Independentei Str., Bucharest, Romania

m.eftimie@gmail.com

Abstract. The mechanical strength of the glasses is determined by the strength of the established bonds between the different macro-anionic species, these bonds being the weaker compared to the covalent ones inside the macro-anionic species. In this paper is proposed a probabilistic relationship to calculate the theoretical mechanical strength of the SiO_2 glass and of some compositions in the $Na_2O{\cdot}SiO_2$ system, in good agreement with literature data.

1. Introduction

A glass, both as a melt and in a solid state, is a system consisting of entities of different types and dimensions (so-called "macro-anionic" species), which can be determined based on some structural models. In a fundamental way, it is considered that the mechanical strength of a material is determined by the bonding forces made between the constituent ions (atoms). For the unary oxide glasses, such as SiO_2 glass, an attempt to calculate the theoretical mechanical resistance, R_t, was based on a Griffith-type equation, obtained from the energy balance related to a Si-O chemical bond break [1], the bond having the equilibrium length $d_e = d$:

$$R_t = \sqrt{\frac{4E\sigma}{\pi d}} \quad (1)$$

where E is the modulus of elasticity; σ - surface energy.

The qualitative and quantitative evaluations that have been made so far in the literature on the structures and properties of the glasses [1-6] allow some clarifications on the way in which the structural particularities of the vitreous state determine the theoretical and practical values of the mechanical strength. A glass, both as a melt and in a solid state, is a system consisting of entities of different types and dimensions (so-called "macro-anionic" species), which gives the structure a nano- and micro-heterogeneous character. Based on some structural models, we can determine the weight (number) of the different macro-anionic species (distribution curve) knowing that between the structural entities the nature of the bonds is predominantly ionic (of the dipole-dipole type), and inside them the bonds are essentially of covalent nature. In this context, the mechanical strength of the glasses is determined by the strength of the bonds that are established between different macro-anionic species, these being the

IOP Conf. Series: Materials Science and Engineering **682** (2019) 012008 doi:10.1088/1757-899X/682/1/012008

weaker compared to the covalent ones inside them. This raises the problem of establishing, qualitatively and quantitatively, the components of these interactions.

2. Mathematical approach

At a first structural level, the glass systems are made up of atoms (ions) of different nature, according to the chemical composition of the glass. This determines, depending on the dimensions of the system, both the number of components and their relative distribution.

A structural image of the $Na_2O \cdot SiO_2$ glass at submicroscopic level is shown in figure 1 [7].

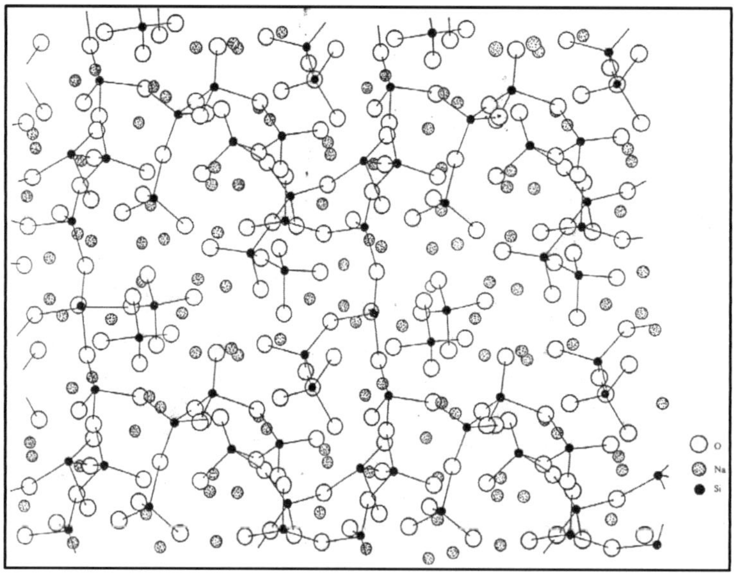

Figure 1. $Na_2O \cdot SiO_2$ glass structure obtained by molecular dynamics simulation [7].

The SiO_2 structure is a three-dimensional one, consisting of [SiO_4] entities linked by corners. The steric characteristics of the structure are represented by the internuclear distance $d_{Si\text{-}O}$, the O-Si-O angle, the angle of Si-O-Si bridges and the type and weight of the cycles formed by several tetrahedrons.

In the case of crystalline SiO_2, for various polymorphic forms, in the ideal situation all these characteristics are strictly constant, in different elemental volumes of SiO_2. Even for real crystals, however, there is some distribution of values of these characteristics around an average value. For instance, in the table 1 [4], for various polymorphic forms of crystalline SiO_2, the range of values in which are placed the actual values for the Si-O internuclear distances, the O-Si-O and Si-O-Si angles (including the average values of them) are presented.

From the data presented in table 1 one can note that:

- even for polymorphic forms of crystalline SiO_2 the steric characteristics present fluctuations around the average values
- the values of the O-Si-O angles have a low variability around the average value of 109.5°, imposed by a sp^3 hybridization, typical for silicon with respect to oxygen
- Si-O-Si angles show a greater dispersion of values around average values
- there is a certain correlation between the internuclear distances Si-O, $d_{(Si\text{-}O)}$ and the Φ = Si-O-Si angles, meaning that for higher values of the Φ angle correspond smaller values for $d_{(Si\text{-}O)}$.

Considering for properties the average values for vitreous SiO_2 ($E = 7 \cdot 10^{10}$ N/m², $\sigma = 0.3$ N/m), for $d = d_e = 1.60 \cdot 10^{-10}$ m, for the theoretical mechanical resistance it results, according to the equation (1), the value $R_t = 1.29 \cdot 10^{10}$ N/m^2 [2]. Assigning reasonable experimental values for the various sizes, a final form for the equation is obtained:

$$R_t = \frac{3.06}{\sqrt{d_e}} \cdot 10^5 \qquad (2)$$

If in equation (2) it is considered - according to [3] - $d = d_e = 1.60 \cdot 10^{-10}$ m, $R_t = 2.42 \cdot 10^{10}$ N/m^2.

Table 1. Si-O internuclear distances, O-Si-O and Si-O-Si angles for various polymorphic forms of crystalline SiO_2

Crystalline structure	d_{Si-O} nm	<d_{Si-O}> nm	O-Si-O (°)	Si-O-Si (°)	<Si-O-Si> (°)
α-quartz	0.1607–0.1611	0.1609	108.7-110.4	143.7	143.7
B-quartz	0.1591–0.1606	0.1598	108.0-110.5	150.9	150.9
α-tridymite monoclinic	0.1576-0.1622	0.1597	-	142.7-179.1	150.0
α-tridymite triclinic	0.1585-0.1623	0.1606	105.8-114.3	139.7-173.2	148.3
B-tridymite	0.5420-0.1563	0.1555	105.9-111.9	165.2-178.7	171.7
α-cristobalite	0,1600-0.1605	0.1602	108.2-111.4	146.8	146.8
B-cristobalite	0.1611	0.1611	107.8-112.8	146.7	146.7
Keatite	0.1562-0.1617	0.1590	103.7-113.8	148.2-159.5	155.2
Coesite	0.1598-0.1631	0.1613	108.3-111.0	136.1-180.0	148.4

Although different calculation methods lead to completely different results [1-3] - depending on the hypotheses of start, the considered simplifications and the actual values due to some physical or steric hindrances - two important conclusions can be drawn for the evaluation of the theoretical mechanical strength of the SiO_2 glass:

- its value is in the range of $1\text{-}3 \cdot 10^{10}$ N/m^2, well above the values recorded in practice, but close to the performances obtained for a series of SiO_2 fibres subjected to special physico-chemical treatments
- whatever the calculation model proposed, in all cases, as well as in equations (1) and (2), a strong dependence between R_t and the Si-O distance results, leading to the general equation:

$$R_t = \frac{C}{\sqrt{d}} \; [N/m^2] \qquad (3)$$

wherein C is a physical constant that can be calculated from theoretical or experimental data, depending on the breakage model considered.

The main idea lies in the fact that the R_t property was usually approached in a deterministic and not probabilistic conception. The probabilistic evaluation mode is supported by both the theoretical structural considerations and the principle equation (3). This is because the internuclear distance does not have a unique value, but has distributed values, even for the SiO_2 glass.

The mechanical strength of the SiO_2 glass has values distributed over a somewhat restricted range of values, $1.78\text{-}1.82 \cdot 10^{10}$ N/m^2, for values of $d_{(Si-O)}$ length in the 0.160-0.166 nm range [2, 8].

The qualitative and quantitative assessments that have been made so far in the literature on the structure and properties of the glass [1-4] allow some clarification on how the structural particularities of the glass state determine the theoretical and practical values of the mechanical strength. This is based on the structural image with the characteristics considered so far, as well as the following:

- the number of coulombian bonds (between the macro-anionic species) formed per unit area, depends on the number (molar weight) of the different macro-anionic species (according to the distribution curve) and on their dimensions. In this way the number of contacts between microparticles with specified dimensions can be determined probabilistically;
- for glasses that are weak bases, macro-anionic species coexist with three-dimensional SiO_2 fragments.

On the other hand, it is expected that, if the SiO_2 weight is higher, the influence of the three-dimensional silica "blocks" on the properties will be prevalent. These components are:

- F_0 attraction force (of a Coulombian nature) between the microparticles;
- the tension due to the different expansion coefficients of the structural entities which are in contact (due to their different chemical compositions);
- the tension caused by different structural defects or/and by some particularities of the microstructure, for example, the separation of the microphases, the existence of superficial microcrystals.

The estimation of the theoretical mechanical strength of the glasses starts from establishing the R_0 value, which, in a first approximation, is proportional to the interaction force F_0 and with n_0, the number of ionic bonds per surface unit. Between the atoms (ions) of silicon, oxygen and sodium, physico-chemical bonds are made, mainly, which are:

- predominantly of covalent type, between silicon and oxygen; these may involve bridged oxygens O_b ($Si\text{-}O_b$) or non-bridged oxygens O_{nb} ($Si\text{-}O_{nb}$);
- predominantly of ionic type, between O_{nb} and sodium ions ($-O_{nb} \ldots Na$);
- dipole-dipole, which are present between the $O_{nb} \ldots Na$ bonds.

Both by their number (relative to the surface unit) and by the value of the interaction force, the first two types of bonds are more important and have a decisive contribution to estimate the average theoretical value of the mechanical strength, $\overline{R_0}$.

For a composition represented by the chemical formula $mNa_2O \cdot SiO_2$, m = 0… 2, if the ratio O/Si is noted with R it results:

- R = O/Si = m + 2
- Number of bridged oxygens: $O_b = 8 - 2R = 4 - 2m$
- Number of non-bridged oxygens (equal to the number of $O_{nb} \ldots Na$ bonds): $O_{nb} = 2R - 4 - 2m = 4 - O_b = 2m$

Similar to SiO_2, it results:
The number of bridged bonds per surface unit and each oxygen, n_1, is

$$n_1 = \frac{4-2m}{4 \cdot V_0^{2/3}} = \frac{1-0.5m}{V_0^{2/3}} \tag{4}$$

analogous, for non-bridged bonds

$$n_2 = \frac{2m}{4 \cdot V_0^{2/3}} = \frac{0.5m}{V_0^{2/3}} \tag{5}$$

In equations (4) and (5), the coordination 4 of Si with respect to O was taken into account and, if V_0 is the volume of the oxygen atom in the glass, the equivalent surface can be considered $V_0^{2/3}$. The volume V_0 can be calculated based on the chemical composition and density, ρ, of the glass, with the equation:

$$V_0 = \frac{M/\rho - (n_{Na} \cdot V_{Na} + n_{Si} \cdot V_{Si})}{n_0} \tag{6}$$

wherein M is the molecular mass of the glass, n and V - the number and the ionic volume, respectively, for the atomic components of the glass.

The force required to break a chemical bond and, consequently, to separate two interacting atoms, F_i, is given by the equation:

$$F_i \approx \frac{E_i}{d_i} = k \cdot \frac{E_i}{d_i} \tag{7}$$

wherein E is the energy consumed to bring the interacting particles to the separation distance d_i (i = 1 for the $Si\text{-}O_b$ bond; i = 2 for the $O_{nb} \ldots Na$ bond); k - constant.

Taking into account the equations 1 to 7, for the average theoretical resistance, $\overline{R_0}$ it follows:

$$\overline{R_0} \approx k\left(\frac{E_0}{d_0}\cdot\frac{1-0.5m}{V_0^{\frac{2}{3}}}+0.5\frac{m\cdot E_2}{d_2\cdot V_0^{2/3}}\right) \qquad (8)$$

Considering that the fundamental equation (7) is valid also for the SiO_2 glass, which has the average theoretical resistance R_0, and based on the equation (8), it results:

$$\overline{R_0} = \frac{R_0\left(\frac{E_0}{d_0}\cdot\frac{1-0.5m}{V_0^{2/3}}+0.5\frac{E_2}{d_2\cdot V_0^{2/3}}m\right)}{\left(\frac{E_0}{d_0\cdot V_0^{2/3}}\right)} \qquad (9)$$

wherein the index 0 refers to the SiO_2 glass.

Table 2. Theoretical mechanical strength of the $mNa_2O\cdot SiO_2$ glass

m	%mol SiO_2	M g/mol	ρ g/cm³	V_0 cm³	d(Si-Ob) nm	$\overline{R_0}\cdot 10^9$ N/m²
0	100	60	2.203	13.6	0.162	42.7
0.25	80	75.5	2.41	13.9	0.163	38.2
0.5	66.7	91	2.51	14.2	0.163	33.7
1	50	122	2.562	16.2	0.164	25.4
2	33.3	184	2.66	16.7	0.167	6.1
2	33.3	184	2.66	16.7	0.167	6.1

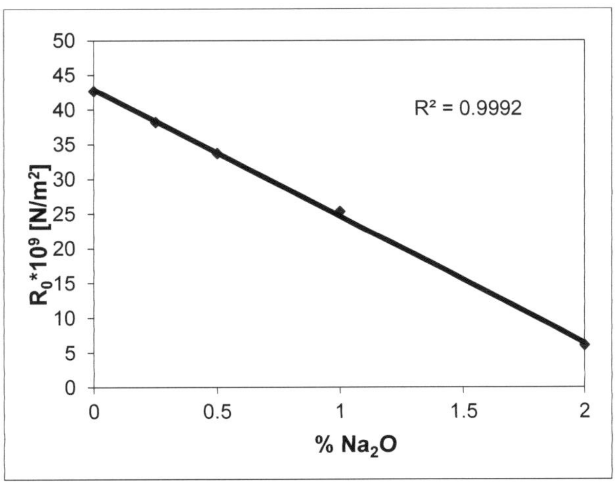

Figure 2. The influence of the Na_2O percentage on the theoretical mechanical strength of the $Na_2O\cdot SiO_2$ glass.

Table 2 presents the calculations for the theoretical mechanical strength and figure 2 shows the influence of the Na_2O percentage on the theoretical mechanical strength of the $Na_2O\cdot SiO_2$ glass.

According to the above statements, we can write a calculus equation of the mechanical resistance R_{0x} for a glass made by macro-anionic species of the $xSiO_2\cdot(2x\text{-}fx)Na_2O$ type.

$$R_{0_x} = F_{0_x} n_{0_x} \qquad (10)$$

Taking into account (8) and considering $W_0 \sim E_0$ it results:

$$R \approx \frac{E_0}{d_0} \cdot \frac{1}{V_0^{2/3}} \quad [N/m^2] \tag{11}$$

But, as the distance between the Si atoms of the two tetrahedra, $d_{(Si\text{-}Si)}$, is calculated, by simple geometrical considerations, with the equation:

$$d(Si\text{-}Si) = 2d(Si\text{-}O) \cdot \sin(\Phi/2) \tag{12}$$

at minimum values of Φ, the distance $d_{(Si\text{-}O)}$ is changed, resulting in the end different values for $d_{(Si\text{-}Si)}$ in the cases of different modes of tetrahedron coupling. Considering the equations (11) and (12), the final calculus equation for R_{0x} is obtained:

$$R_{0_x} = R\frac{E_x}{E_0} \cdot \frac{d_0}{d_x} \cdot \frac{V_0^{2/3}}{V_{0x}^{2/3}}\left(1 - \frac{f_x}{2x}\right) \quad [N/m^2] \tag{13}$$

A check of the correctness of the equation (13) can be done considering the experimental data provided by the literature. Thus, for the activation energy of the viscous flow of SiO_2, quite scattered values are reported in the literature, between 443 kJ/mol and 669 kJ/mol. [1] For the glasses of the Na_2O-SiO_2 system the values are between 63 kJ/mol (more basic glasses) and 125 kJ/mol (more acid glasses) [1].

In relation (13) d_0 is basically the Si-O internuclear distance in the SiO_2 glass, so $d_0 = 0.162$ nm. For the atomic volume of oxygen V_0 in the SiO_2 glass and V_x in glasses of the Na_2O-SiO_2 system, the following assessments can be made: $V_0 = 13.6$ cm^3, and V_x varies between 13.8 and 15.1 cm^3 [1].

The calculation of some properties of the silicate glasses based on a macro-anions theory for different chemical compositions showed good agreement with the experimental data in the range of 50-80% mol SiO_2, after which some systematic deviations between the calculated and experimental values are recorded [2].

The next step of the calculation requires the determination of the number of bonds between the various macro-anionic species *i* and *j*, as well as of the values σ_{ij}.

The tension σ that appears between two macro-anionic species with different expansion coefficients is given by the known equation:

$$\sigma_{12} = \frac{(\alpha_1 - \alpha_2) \cdot E \cdot (T_g - T_0)}{1 - \mu} \quad [N/m^2] \tag{14}$$

in which: α_i - the average expansion coefficient of microparticle *i*, determined over the temperature range T_0 - T_g [grd^{-1}]; E - Young's module [N/m^2]; T_g - max (T_{g1}, T_{g2}), T_{gi} - glass transition temperature of *i* glass [grd]; T_0 - ambient temperature [grd]; μ - Poisson's coefficient.

A macro-anionic species *i* may come into contact with another macro-anionic species *i* (the total number of these contacts being $N_{ii} = C_{ni}^2$) or with a macro-anionic species *j* (the number of these pairs being $N_{ij} = C_{ni}^1 \cdot C_{nj}^1$), where C_m^p represents the combinations of *m* objects taken as *p*. The total number of N_t interactions is given by the equation:

$$N = \sum_{i=1}^{N} N_{ii} + \sum_{i=1}^{N} \quad \sum_{j-2}^{N-1} C_{ni}^1 \cdot C_{nj}^1 = C_N^2 = N(N-1)/2 \tag{15}$$

From here, the frequency of occurrence *f* of different interactions between entities of the same type and of different types is, respectively:

$$f_{ii} = \frac{C_{ni}^2}{C_N^2} = \frac{n_i(n_i - 1)}{N(N-1)} \tag{16}$$

$$f_{ij} = \frac{C_{ni}^1 \cdot C_{nj}^1}{C_N^2} - = \frac{2n_i \cdot n_j}{N(N-1)} \tag{17}$$

For large values of *N*, the calculation of the values of f_{ii}, f_{ij}, α_{ij}, σ_{ij} is particularly laborious. For this reason, we illustrate here an application for a glass of sodium metasilicate composition for which the first 16 macro-anionic species are considered. The values for n_i and n_j are given by the molar percent p_x

for the entities with x silicon atoms, and $N = 100$. Following the calculations, one can observe that two distinct situations arise, in which the actual resistance of the glasses is differently determined:

- In a first situation, in which the frequency of occurrence of the different values for R_0 is made as a histogram, the values recorded for R_0 are always higher than the maximum stresses that appear between different macro-anionic species. In this case, the actual mechanical strength of the $R_0^{(1)}$ glass is:

$$R_0^{(1)} = R_{0_{min}} - \sigma_{max} \quad (18)$$

- A second possibility is to obtain a graph with the radial distribution of the theoretical mechanical strength of the glasses (surface with micro-cracks) [9]. It is observed that, in this case, a series of values of σ exceed some values of R_0, which is equivalent to the breaking of bonds with the appearance of some cracks. Because the breakage of bonds occurs in the contact area of two microparticles, the size of the microcracks can be assimilated to the length of a circle arc (if we consider spherical microparticles) that can reach the maximum value of π^D (D - diameter of the macro-anionic species). In this situation the actual mechanical resistance $R_0^{(2)}$ of the glasses is calculated with Griffith's equation:

$$R_0^{(2)} = K\sqrt{\frac{E\cdot\gamma}{D}} \; [N/m^2] \quad (19)$$

in which: K - factor that depends on the geometry of the microcracks; E - modulus of elasticity [N/m^2]; γ - surface tension [N/m]; D - characteristic dimension of the crack [m].

To check the relationships deduced from theoretical structural considerations, an application will be made for the case of a sodium metasilicate glass. The number of macro-anionic species considered for sodium metasilicate glass is N = 16, in which case their weight is over 0.7%. For this case, the situation in which $\sigma_{max} < R_{0min}$ is more probable. In this way, the mechanical resistance $R_0^{(1)}$ is:

$$R_0^{(1)} = 98\cdot 10^6 \text{ N/m}^2$$

If, however, in the case of the glass of the same composition, about 40 macro-anionic species are taken into account - when the last macro-anionic species is present with a weight of about 0.1% - for R_0, the value of $68{,}7\cdot 10^6$ N/m^2 results, $\Delta\alpha_{ij} = 310\cdot 10^{-7}$ grd^{-1}, and $\sigma_{ij} = 88.3\cdot 10^6$ N/m^2. Because $\sigma_{max} > R_{0min}$, microcracks may occur due to breaking bonds. In the case of sodium metasilicate, the first 16 macro-anionic species have values between 0.5 and 1.2 nm [2], but when x (the silicon atoms) increases, the diameter of the macro-anionic species increases to tens or even hundreds of nm and, therefore, in the same way but three times more, increases the length of the micro-cracks, this result being in accordance with experimental determinations. Table 3 shows for different values of D [nm], the mechanical strength of the glasses, according to Griffith's equation.

Table 3. The mechanical strength of the glass for different values of D

Microcrack dimension nm	0.162	5	10	50	150
mechanical strength $\cdot 10^{-7}$ N/m^2	216	19.4	13.7	6.1	3.5

We can observe from table 3 a somewhat surprising result that has been observed experimentally, also, [10] and is theoretically explainable through the statements made above: with the increase of the percentage of SiO_2 a decrease of the mechanical resistance takes place.

IOP Conf. Series: Materials Science and Engineering **682** (2019) 012008 doi:10.1088/1757-899X/682/1/012008

3. Conclusions

In this paper, the theoretical mechanical strength for $Na_2O \cdot SiO_2$ and SiO_2 glasses was calculated, and the results were compared with the experimental values provided by the literature. The calculation of some properties of silicate glasses based on a macro-anions theory for different chemical compositions showed good agreement with the experimental data in the 50-80% mol. SiO_2 range, after which some systematic deviations between the calculated and the experimental values are recorded. In the case of sodium metasilicate, following the theoretical calculations it was observed that with the increase of the percentage of SiO_2 there is a decrease of the mechanical resistance, observed in experimental results, too.

Although mechanical resistance has been studied for so many years, it remains topical because the classical approaches lead to unique mean values, by considering the SiO_4 a perfect tetrahedron in respect to the internuclear angles and distances. Regardless of the calculation method used, the same conclusions are reached: there is a strong dependence between R_t and the distance Si-O, d_e. It turns out that mechanical resistance must follow a probabilistic approach, taking into account the distribution of angles, chemical bonds, structural entities, internuclear distances that may have values with a certain probability (frequency of occurrence). This reality confirms the distributed mechanical strength values measured for the glasses.

4. References

[1] Scholze H 1990 *Le Verre* (Paris: Dunod)
[2] Balta P 1984 *Tehnologia sticlei* (Bucharest: Editura Didactică și Pedagogică)
[3] Naray-Szabo I and Ladik J 1960 *Nature* **188** 226
[4] Hubner K 1977 *Phys. Stat. Sol.* **(a) 40** 487
[5] Varshneya A 2006 *Fundamentals of Inorganic Glasses* (Sheffield: Society of Glass Technology)
[6] Lehman R 2017 *The Mechanical Properties of Glass*, *Glass Engineering 150:312* (New Jersey) available at http://glassproperties.com/references/MechPropHandouts.pdf
[7] Inoue H and Yasui I 1987 *Phys. Chem. Glasses* **28** 2
[8] Pauling L 1980 *Am. Mineral* **65** 321
[9] Radu D 1982 *Rev. Chim.* **8** 726
[10] Liebau F. 1985 *Structural chemistry of silicates: structure, bonding, and classification* (Berlin, New York: Ed. Springer-Verlag)

IOP Conf. Series: Materials Science and Engineering **682** (2019) 012009 doi:10.1088/1757-899X/682/1/012009

Experimental analysis of AZ31B magnesium alloy sheet failure using punch stretching

C Grigoras, B Chirita, G Brabie and C Ciofu

"Vasile Alecsandri" University of Bacău, Industrial Engineering Department, Calea Mărăşeşti Street No. 157, Bacău, România

grigorascosmin_phd@yahoo.com

Abstract. The formability of magnesium alloy sheets at room temperature presents anisotropy in mechanical properties and difficulties in terms of occuring cracks easily, especially in regions with bend radius. In addition, the elastic spring-back is significant, leading to massive deviations from the desired shape. Recent studies conducted in this field lead to stretch forming magnesium alloys sheets using thermo-mechanical treatments at temperatures up to 400°C. The present study was conducted on 1 [mm] thick magnesium alloy AZ31B sheet to investigate its formability when stretching at room temperature by several dies with different radius. The stretching process was conducted on a hydraulic press, using 3D printed PLA dies with the following values: R180, R320, R540, R720, R900 and R1080. The samples were stretched until fracture to highlight the fracture force, distance to fracture, deviation from the die radius and bend angle.

1. Introduction

In the last decade, the interest has increased significantly on magnesium-based alloys, especially in the field of automotive industry. The high interest is due to the fact that magnesium alloy have a very good strength to weight ratio, high vibration damping and effective electromagnetic screening [1]. Contrariwise the downside of this alloys is that, at room temperature, they present low mechanical strength, poor microstructure control during processing and anisotropy of fatigue [2], [3]. There is an undesired failure mode, usually called the shear fracture, which mainly results from the low ductility of advanced metal sheets. Metal sheets with little amount of necking exhibit the fracture surface slanted along the maximum shear stress direction through its thickness [4]. In traditional thermo-mechanical processing of magnesium alloy sheets (such as extrusion) there are always obtained strong basal textures, with the c-axis of most grains being nearly parallel to the normal direction (ND) of sheet. These Mg alloy sheets usually exhibit low plasticity and formability at room temperature, which lead recent researches to focus on improving texture by tailoring. It has been accepted that the press formability of sheets at room temperature is strongly affected by the Lankford value (r-value) [5]. Researchers proposed to reduce the springback in bending of ultra-high strength steel sheets by bottoming and additional bending. Also, lubricants, coatings on the tool surface and tool material are effective to improve the tool wear. However, the reduction in shearing force by control of deforming behavior of the sheet in punching and shearing is an alternative approach [6]. When sheet metal forming magnesium alloys the deformation is defined by the flow curves. Regardless of the type of forming at room temperature classic processes do not offer the required results, leading to prematurely breaking [7].

IOP Conf. Series: Materials Science and Engineering **682** (2019) 012009 doi:10.1088/1757-899X/682/1/012009

2. Experimental set-up

This experimental work is part of a comprehensive case study. The purpose is to determine how magnesium alloys sheets behave when using stretching and single point incremental forming (SPIF) with different types of dies and working temperatures. The present paper aims to study the behavior of magnesium alloy AZ31B sheets when stretch to the breaking limit. This study aims to analyze the force and distance required to fracture, and the spring-back effect for both radius and bend angle. The results are necessary to develope a system that allows to perform stretch forming and SPIF processes using the same clamping device. The experimental setup for this research paper is composed of two parts: the 3D printing process and the stretching process setup.

2.1. 3D printed die

Rapid improvement in 3D printing technologies has led to substantial improvement in the complexity of shape, size, and materials that can be obtained and used. From simple home-made parts to entire industries (aerospace, automotive, organ replication, civil constructions), it is expected that, in the next decade, this process to be the next key manufacturing technology [8]–[10]. Our approach is to use this technology for 3D printing forming punches, figure 1. This comes as a solution to the problem generated by the complex machining procedures that take place when manufacturing from metal.

A total number of 6 punches where printed, with various radii (180, 360, 540, 720, 900, 1080). They were manufactured on an Ultimaker Original Plus 3D printer, using 2.85 [mm] thick Innofill PLA wire. The printing parameters taken into account were: layer height (0.4 mm), wall thickness (1.6 mm), infill density (75 %), infill pattern (45° lines), printing temperature (210 C°), build plate temperature (60° C) and printing speed (60 mm/s). An auxiliary fastening system was manufactured in order to assure the connection between the dynamometer and the hydraulic press (figure 1).

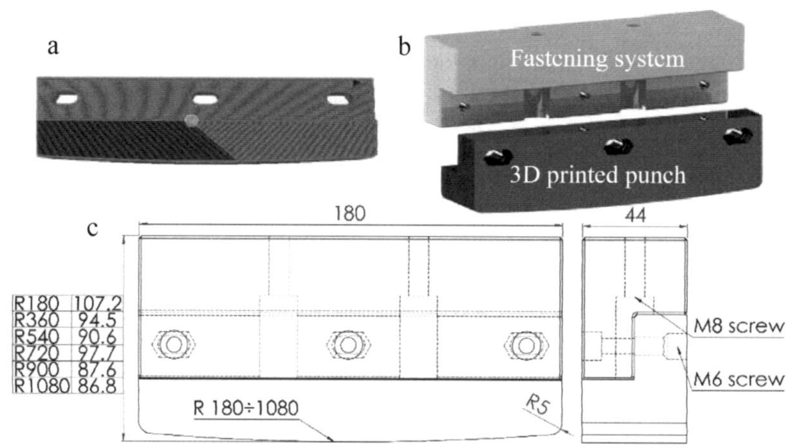

Figure 1. a) 3D printed trajectories b) fastening system and c) assembly drawing

3D printed parts deviate from the desired shape, as the plastic material cools. As a result the punches were measured for radius size and deviation from circularity on a 3 axes Tesa 3D Micro Hide CMM equipment. The designed shape and the measured values are presented in table 1. As can be noted, there are deviations from the ideal shape so these measurements were taken as reference.

Table 1. 3D printed punches measurements

Designed radius [mm]	180	360	540	720	900	1080
Measured radius [mm]	182.383	365.146	553.904	738.847	942.176	1149.946
Deviation from circularity [μm]	0.005	0.123	0.006	0.06	0.061	0.083

IOP Conf. Series: Materials Science and Engineering **682** (2019) 012009 doi:10.1088/1757-899X/682/1/012009

2.2. Set-up description

The experimental study was conducted using 1 [mm] thick magnesium alooy AZ31B blanks, cut at 315 x 46 [mm]. The equipments used was a hydraulic press (Hydramold HPHL-075.300), on which can be controlled the travel distance and speed of the hydraulic driving system (1) in the vertical direction (figure 2). The base plate (2) has two roles: it fixes the hydraulic clamping system (4) and, because it has a central cut up, allows the punch free travel. The blank (5) is fixed in position using metallic jaws (3), on which the hydraulic clamping system operates at 20 [bar]. The punch (6) is fixed to the Kistler 9272 dynamometer (8), using the fastening system (7).

The experimental plan is based on three deformation speeds (0.2, 0.7, 1.2 [mm/s]), for each punch, meaning a total number of 18 trials. The study implies that the magnesium blanks are stretch until failure, at room temperature. The values for force and distance were obtained from the dynamometer's software (DEWESoft X3), while the shape measurements were made using the GOM ATOS 3D imaging metrology system.

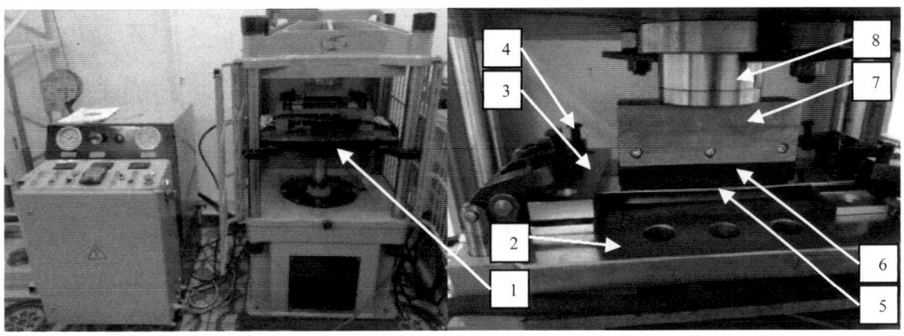

Figure 2. Experimental setup of the stretch failure process
1-hydraulic driving system, 2-base plate, 3-jaws, 4-hydraulic clamping system, 5-AZ31B blank, 6-3D printed punch, 7-fastening system, 8-dynamometer

Shown in figure 3 is the failure limits mechanism of the magnesium blanks. There can be observed three distinct steps (4, 5 and 6), which coincide with the initial position (4-the punch touches the blank), the intermediary position (5) that is right before breaking and the finish position, which represents the distance to fracture. After breaking (6) and unclamping the parts presented a major spring-back effect. Coresponding to the obtained parts the fallowing parameters were studied: R_1 (the radius that corresponds to the punch radius) and θ_1 (spring-back angle).

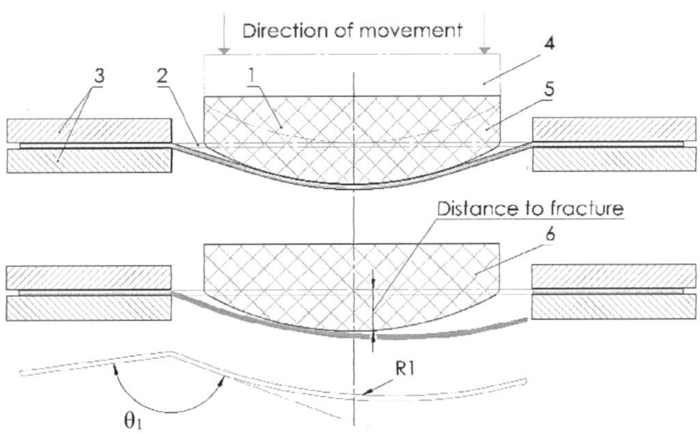

Figure 3. Schematic representation of the stretching process
1-Punch, 2-AZ31B blank, 3-jaws, 4-initial position, 5-intermediary position, 6-finish position

IOP Conf. Series: Materials Science and Engineering **682** (2019) 012009 doi:10.1088/1757-899X/682/1/012009

3. Results and discussions

The input parameters were the deformation speed and the punch radius, while for the results were: fracture force, distance to fracture, bend angle and radius. As the blanks were stretch until fracture, the ideal shape was considered position 5, figure 3. Compared to the ideal shape, the magnesium AZ31B blanks present significant deformations (figure 4). The elastic spring-back, in this case, was analyzed for R1 and θ_1, figure 3.

R180 part sample

R1080 part sample

Figure 4. Stretched parts and 3D scanned model

3.1. The analysis of the fracture force and distance to fracture

Highlighted in figure 5 is the distribution of the force with reference to the deformation speed. A slight increase can be noted when using higher speeds. Although the maximum recorded force of 3163.6 [N] was identified when stretching with 0.2 [mm/s], the overall values are higher for 0.7 and 1.2 [mm/s]. The minimum amount of force required to break was 1680.21 [N]. It can be noted that failure has occurred under 2000 [N], only at slower speeds.

In the case of the punch radius effect, presented in figure 6, the maximum force increases as the punch radius has higher values. A lower fracture limit was recorded when R360 and R540 punches were used, while the R720 and R1080 ones record the highest value.

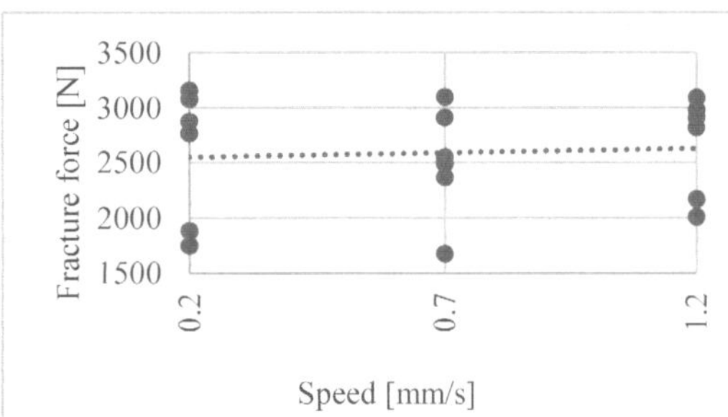

Figure 5. Influence of the speed [mm/s] on the fracture force [N]

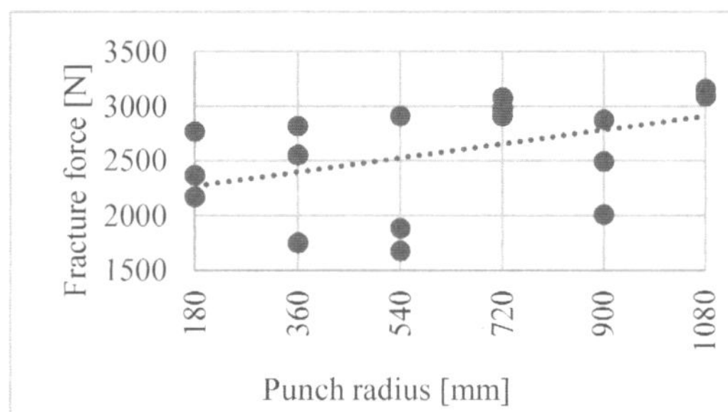

Figure 6. Influence of the punch radius [mm] on the fracture force [N]

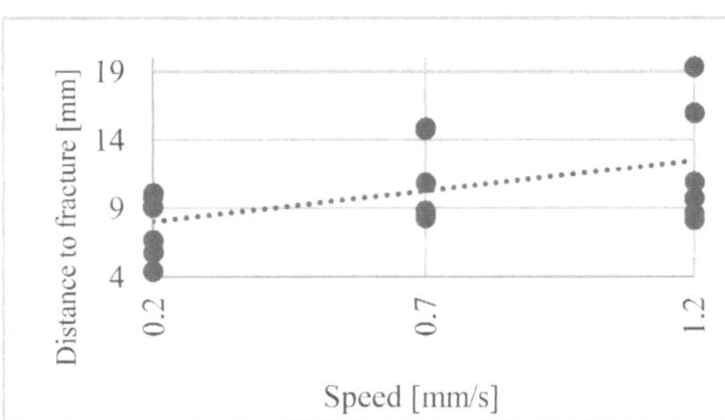

Figure 7. Influence of the speed [mm/s] on the distance to fracture [mm]

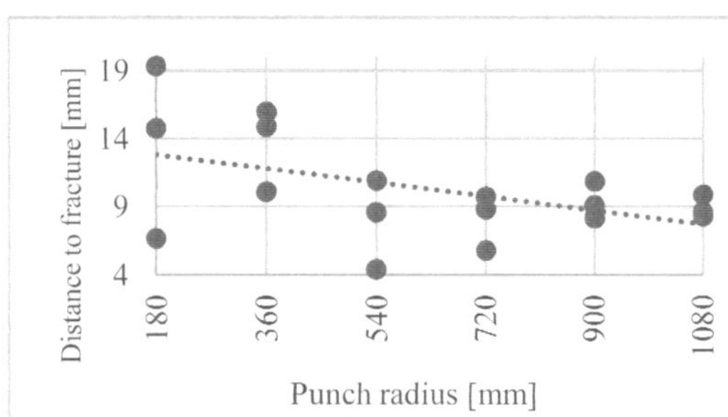

Figure 8. Influence of the punch radius [mm] on the distance to fracture [mm]

The influence of the speed and punch radius upon the distance to fracture can be noted in figure 7 and figure 8. The distance to fracture presents higher values when higher speeds are used. The values are distributed between 8 and 12 [mm]. Below this threshold only the 0.2 [mm/s] speed has recorded values. A higher breaking limit is recorded when higher speed are used, 14.77 [mm] for 0.7 [mm/s], respectively 15.96 and 19.32 for 1.2 [mm/s]. The distance to fracture presents greater limits when smaller radius punches were used. Although the parts stretched with the R1080 punches withstand higher forces they have a lower distance to fracture. There can be noted that using R180 and R360 punches leads to a higher distance to fracture, over 14 [mm].

3.2. The analysis of the punch radius deviation, bend radius and angle

This analysis was conducted taking into consideration the spring-back coefficient (1), (2) for the obtained punch radius and bend angle. A factor close to 1 means that the spring-back effect is weak, while when is near to 0 heavy significant deformation occurs. For the angle spring-back the coefficient takes into consideration only the initial and final measurements whereas the one for the radius is defined regarding the neutral deformed fiber. The theoretical angle was determined analytically from position 5 (figure 3), witch correspondes to the position of a tangent line connecting the jaws and the punch radius against the initial blank position. When analisyng the radius spring-back the assumption was that the required radius of the part is that of the corresponding punch.

$$K_\theta = \frac{\theta}{\theta_1} \tag{1}$$

$$K_R = \frac{R + 0.5t}{R_1 + 0.5t} \tag{2}$$

Represented in figure 9 is the effect of the theoretical and measured angle on the spring-back coefficient K_θ. The results have shown that the difference in initial and final angle is between 2.39 [°] and 28.96 [°]. A large distribution of the measured values is between 150 and 160 [°] while the theoretical ones are from 166 to 178 [°]. This difference in results leads to a sping-back coefficient value from 0.83 to 0.98.

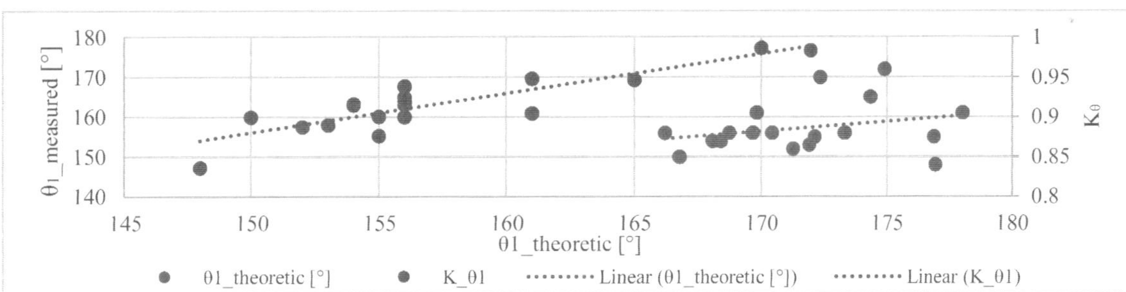

Figure 9. Effect of the theoretical and measured angle on the spring-back coefficient K_θ

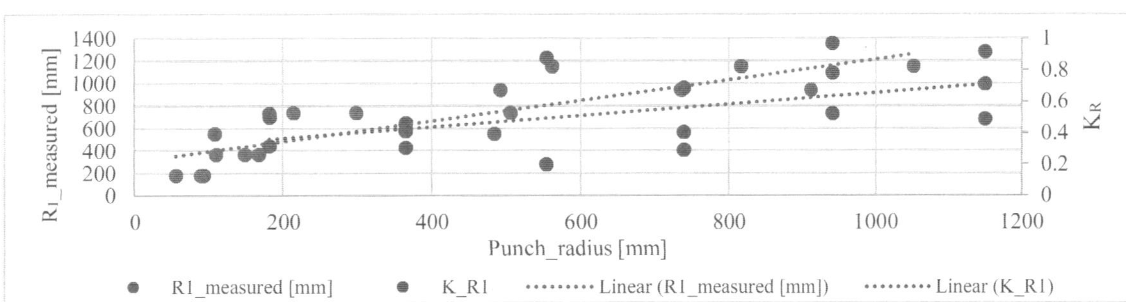

Figure 10. Effect of the punch radius and blank radius on the spring-back coefficient K_R

The radius spring-back coefficient values are spreaded over a larger field, with values between 0.19÷0.98 (figure 10). Most of the obtained values are lower than 0.5, witch are specific to severe deformation. These results are obtained when small radius punches are used. In almost 25% of the cases the spring-back effect ranges between 0.19÷0.30.

4. Conclusions

The present study aimed to highlight the fracture limits when stretching magnesium alloy AZ31B sheet over a 3D printed die by analising the maximum force, distance to fracture, spring-back angle and radius.

It has been concluded that the fracture force depends on the punch radius, as the contact surface increases, therefore the contact pressure. With the increase in force the distance to fracture decreases. This decrease was also observed when using large radius punches.

The spring-back effect indicates significant deformations in the case of the radius, particulary when small radius punches are used. In the case of the sping-back angle the experimental study indicates that severe deformation take place

AZ31B magnesium sheets present low resistance with high deformations when stretch. This phenomen thakes place as magnesium alloys have low ductility at room temperature. Stretch forming can be carried out when higher temperatures are used.

The 3D printed punches were not affected during the stretch process.

5. References

[1] Liu D, Liu Z and Wang L 2014 Simulation of Rolling Process of AZ31 Magnesium Alloy Sheet Procedia Eng. 81 173–178

[2] Miao Q, Hu L, Wang G and Wang E 2011 Fabrication of excellent mechanical properties AZ31 magnesium alloy sheets by conventional rolling and subsequent annealing *Mater. Sci. Eng. A* **528(22-23)** 6694–6701

[3] Rao K P, Prasad Y V R K and Suresh K 2011 Materials modeling and simulation of isothermal forging of rolled AZ31B magnesium alloy: Anisotropy of flow *Mater. Des.* **32(5)** 2545–2553

[4] Park N, Huh H and Yoon J W 2018 Anisotropic fracture forming limit diagram considering non-directionality of the equi-biaxial fracture strain *Int. J. Solids Struct.* **151** 181–194

[5] Shi B Q et al. 2019 Achieving ultra-low planar anisotropy and high stretch formability in a Mg-1.1Zn-0.76Y-0.56Zr sheet by texture tailoring via final-pass heavy reduction rolling *Mater. Sci. Eng. A.* **746** 115–126

[6] Yonekawa R and Sedoguchi K 2018 Shearing of ultra-high strength steel sheets with step punch *Procedia Manuf.* **15** 597–604

[7] Friedrich H E and Mordike B L 2006 Magnesium Technology (Berlin/Heidelberg: Springer-Verlag)

[8] Grenda E 2019 *Additive Fabrication Spy Sampler* (Castle Island Co.)

[9] Marcovitz H 2016 *What Is the Future of 3d Printing? (The Future of Technology)* (Referencepoint Pr. Inc.)

[10] Winnan C D 2013 *3D Printing: The Next Technology Gold Rush - Future Factories and How to* Capitalize on Distributed Manufacturing (Create Space Independent Publishing Platform) p 290

Acknowledgments

This work was supported by a grant of the Romanian Ministry of Research and Innovation, CCCDI – UEFISCDI, project number PN-III-P1-1.2-PCCDI-20170446 / 82PCCDI / 2018, within PNCDI III.

IOP Conf. Series: Materials Science and Engineering **682** (2019) 012010 doi:10.1088/1757-899X/682/1/012010

Applications of piezoelectric MFC type transducers in mechatronic measurement systems - the impact of a mechanical subsystem damage on a system operation

M Płaczek

Silesian University of Technology, Faculty of Mechanical Engineering,
Institute of Engineering Processes Automation and Integrated Manufacturing Systems,
Konarskiego 18A, 44-100 Gliwice, Poland

marek.placzek@polsl.pl

Abstract. The work include a report of research works concerning applications of non-classical piezoelectric transducers in mechatronic measuring systems. Composite piezoelectric transducers called Macro Fiber Composite (MFC) are considered. Characteristics of considered transducers were made, indicating their advantages and disadvantages, as well as the results of carried out laboratory tests are presented, in which MFC transducers were used as sensors for mechatronic measurement systems. A series of laboratory tests was carried out on the created laboratory stand to determine the impact on the measured values of damage of the mechanical subsystem on which the MFC-type piezoelectric transducer was glued. Results of measurements are presented and analysed in relation to different applications of the piezoelectric MFC sensors.

1. Introduction

Piezoelectric materials are finding an increasing number of applications in modern technical devices due to the number of advantages that this type of intelligent materials is characterized by. One of these advantages is the ability to directly convert both electricity into mechanical work by a piezoelectric transducer powered by alternating voltage (inverse piezoelectric effect), and the ability to generate an electric charge under the influence of mechanical loads to which the transducer will be subjected (simple piezoelectric effect). For this reason, piezoelectric transducers can be used both as actuators and driving elements, as well as sensors. The progressive development of materials with piezoelectric properties is aimed at improving their efficiency of energy transformations underlying the operation of piezoelectric transducers. Starting from the classic ceramic transducers, e.g. PZT type, where the disadvantages limiting the field of their potential applications are the fragility of the transducers and their large mass, in the course of research and scientific work, flexible and highly efficient composite transducers were developed. These types of transducers include piezoelectric films. An example of this was the development of composite piezoelectric transducers. In 1996, the American Langley Research Center invented cheap, innovative piezoelectric composite transducers designed, among others, to control vibrations, noise and deviations in structural composites of beams and panels, as well as to recover electrical energy from mechanical vibrations. The developed piezoelectric transducers are called Macro Fiber Composite (MFC) and they have the form of a thin, flexible piezoelectric film. In 2002, MFC transducers were commercialized as a licensed and patented invention, which is constantly improved and adapted to the specific needs of customers and new applications. NASA initially developed piezoelectric film to develop and improve aircraft structures by using MFC film in aircraft wings and helicopter rotor blades. MFC piezoelectric film consists of rectangular piezoelectric ceramic rods, placed between polymer layers containing electrodes that transfer voltage directly to piezoelectric rods (their thickness does not exceed a few tenths of a millimetre) [1,2]. As a result of powering the piezoelectric transducer from an external source, the following: tensile, bending or torsion of the transducer is possible. It is also possible to use MFC films in systems for recovering electrical energy from mechanical vibrations [3,4].

As a part of the author's research on MFC piezoelectric transducers, issues of modelling and application using both simple and inverse piezoelectric effect were discussed. This paper presents selected proposals for the application of MFC transducers in mechatronic measuring systems, as well as the limitations and disadvantages of such systems that should be considered during their design. Also results of a series of laboratory tests that were carried out to determine the impact on the measured values of damage of the mechanical subsystem on which the MFC-type piezoelectric transducer was glued are presented. Mechatronic systems are characterized by a synergistic combination primarily of mechanical, electrical and IT engineering, which leads to the development of modern technical means [5,6]. This article, in chapters 2 and 3, presents a review of the subject literature and a synthetic description of previously published research results, which aims to introduce the reader to the most important elements of the considered issues. Chapter 4 presents, however, unpublished results of preliminary research on the analysis of the impact of mechanical subsystem damage in the proposed measuring system based on the use of MFC piezoelectric transducers on the effectiveness of its operation. The thesis was summarized by drawing conclusions based on both, previously published studies and the results of the latest research.

2. Applications of MFC transducers in mechatronic measurement systems

In the research and development project regarding the modernization of the body of freight wagons, the idea of operation of the technical condition supervision system for modernized freight wagons was proposed. The paper [1] presents issues regarding the use of piezoelectric films to monitor the technical condition of freight wagons during their operation. The results of laboratory tests that were carried out on a freight car model created on a scale were presented. The possibility of detecting changes in the technical condition of the wagon based on the analysis of its dynamic response was determined. The research consisted in measuring the dynamic response of the model while driving. To measure the vibrations of the supporting structure of the model, a piezoelectric Macro Fiber Composite (MFC) transducer was used, which was glued on its surface. A series of tests was carried out on the model without load and with load, as well as with obstacles located on one or both rails. The waveforms obtained were analysed. The conducted works were aimed at verifying the possibility of inferring about the state of the examined object based on measuring its response to dynamic excitations and the possibility of its separation from noise, which is the result of its work in normal operating conditions. It was shown that it is possible to infer about the technical condition of the tested object based on signals generated by properly placed piezoelectric sensors, affixed on the wagon components. As a part of the work, measurement points on the created freight wagon model were selected, and a series of tests was carried out. Tests carried out on the created laboratory stand showed the effectiveness of the proposed method. It is possible to analyse and interpret the electric voltage signal generated by piezoelectric film stuck on the surface of the tested object in order to infer its condition. Papers [7-8] present the results of tests and verification of system assumptions carried out on real objects. During these tests, MFC transducers were glued to structural elements of a standard freight wagon and the voltage waveforms generated by them during the observed rides in the conditions of operation of the wagon close to standard operation were recorded. At the current stage of research, the recorded waveforms were used to excite vibrations of elements of a modernized freight wagon box equipped with composite panels. By measuring the dynamic response to given excitation, the possibility of detecting the presence or absence of elements of the modernized plating was determined. The subject of further work will be the

IOP Conf. Series: Materials Science and Engineering **682** (2019) 012010 doi:10.1088/1757-899X/682/1/012010

construction of a prototype freight wagon with modernized shell plating and conducting long-term operational tests of the plating damage detection system.

Systems based on piezoelectric transducers applications are also used in railway transport and issues related to the study of the dynamic response of both railway rolling stock and railway infrastructure are the subject undertaken by a lot of researchers. Dynamic loads generated during the passage of railway trains are deleterious to the infrastructure, such as bridges or buildings located near tracks, as well as people staying in them. There is however the possibility of their use in systems for condition monitoring of infrastructure components [9-14]. Also important are issues of protection of building against vibration generated by the ground and underground rail transportation as well as protection and minimizing the impact of vibrations on both passengers and persons in the buildings surrounding the railway line [10].

An interesting idea of piezoelectric films application is also an innovative system for traffic supervision and detecting and alerting about collisions presented in paper [15]. The purpose of the research was to conduct theoretical and experimental research in the field of analysing the dynamic response of road infrastructure elements, such as protective barriers, soundproof screens, lighting elements, etc., to extortion caused by ongoing traffic. The implementation of research tasks was aimed at acquiring new knowledge in the field of phenomena occurring during the propagation of vibration energy, generated by passing vehicles, on infrastructure elements, and the possibility of measuring the dynamic response of these elements using glued MFC piezoelectric foils, as well as analysing the received signals for their use in modern surveillance systems.

There are known methods for measuring vehicle speed, traffic intensity and controlling traffic lights, implemented through embedded sensors in the form of induction loops or piezoelectric sensors, which also allow measuring the weight of passing cars. Such solutions, however, require interference with the surface, which makes it impossible to install them without impeding traffic or incurring high costs. In the proposed solution it is possible to carry out this type of tasks by measuring the dynamic response of road infrastructure elements located directly in the vicinity of the road gauge. To this end, experimental research was conducted, in which the hypothesis was verified. The measurements of the dynamic response of protective barriers were carried out on the highway. The sensors were glued on the elements of the protective barrier that separates the lanes in opposite directions and is located on green belt. In order to perform the measurements a mobile voltage signal acquisition system station was constructed. To build up the system and for the data acquisition equipment and software from National Instruments were used. Four of the MFC piezoelectric films were connected with the measuring card NI-9215. Then the measuring card was installed in the module cDAQ-9191. The measurements were carried out simultaneously with the recording of the passing vehicles with the camera. Recorded video was then compared with the voltage waveforms generated by the piezoelectric transducers in order to analyse and interpret the results.

By analysis of signals received during each of the tests, it was found that the profile belt is too susceptible element. Numerous spikes of the generated signal voltage were observed reflecting not passing of any vehicle. These signals resulted from weather conditions (wind) and were not results of a vehicle movement. Most precisely the vehicle passing is reflected by the waveform generated by the piezoelectric film mounted on the pillar. Passing of a vehicle of weight over 3,5 tons moving along right and left lanes of the road as well as passing of a vehicle of weight up to 3,5 tons moving along the left lane of the road can be easily detected. Signals generated by all transducers were also affected by vehicles travelling in the opposite direction on the road lanes separated by a green belt.

An important advantage of this kind of system is the assembly of its components to existing elements without violating the road surface and the retention of the traffic. The system will also have a very low demand for electricity, which also can be met using renewable energy sources such as solar, wind and even recovery of electric energy from mechanical vibrations using piezoelectric elements making up the system. On the other hand, there are also a few problems that must be solved. The impact of the vehicles going along the opposite road lane as well as the possibility to detect light vehicles driving on the right lane can be solved by a proper distribution of sensors. In the presented work sensors were glued on the elements of the protective barrier that separate the lanes in opposite directions and is located on green belt. The mentioned problems should be solved by distribution of sensors on all protective barriers (external barriers and barriers on the green belt). Waveforms generated by all sensors in the system can be then filtered and juxtaposed in order to correctly infer the type, location and direction of movement of vehicles. It will be the aim of the author's future work.

3. Reliability of measuring systems based on the use of MFC piezoelectric transducers

When developing the concept and designing mechatronic systems in which MFC piezoelectric transducers are used as sensors, the possibility of disturbing the measurements by various external factors should be considered. These may include fluctuations in the system's operating temperature, electromagnetic disturbances or improper performance or damage to the layer connecting the transducer to the mechanical subsystem [16]. For this reason, studies about the impact of electromagnetic disturbances as well as changes in their operating temperature on the effectiveness of MFC piezoelectric transducers were carried out [17-20].

During the study of the impact of electromagnetic disturbances presented in paper [17] during laboratory tests two Macro Fiber Composite (MFC) piezoelectric transducers were used. Both were glued on the surface of the cantilever beam made of aluminium alloy. The first one was used for vibration excitation. It was connected to the high voltage amplifier supplied by the wave generator. Harmonic electric voltage was applied to the transducer to generate harmonic excitation of the beam. The beam was excited by the MFC transducer with frequency equal to 7,1 Hz. It was the first natural frequency of the beam. In the case of excitation with the first natural frequency it was possible to obtain the highest values of the beam deflection amplitude. It was very important to obtain the maximum value of the electric voltage generated by the second MFC transducer which was used as a sensor of the beam's deflection. According to the beam harmonic vibrations the MFC sensor generates harmonic electric voltage. This electric voltage was measured using NI 9191 CompactDAQ Chassis equipped with C series voltage input module. During the tests, the influence of conducted disturbances on the transmission wire between sensor and the measure module was examined. The EM-clamp consisting of a tube of split ferrite rings of two different grades, which can be clamped over the cable under test in a non-invasive way was used in order to introduce disturbances into the system. The MFC sensor was connected to the measuring equipment using shielded wire but during the measurements the shield was connected as follows:

- disconnected on the both ends of the wire,
- connected only to the MFC transducer (to the negative potential wire), while the second end of the shield was disconnected,
- connected only to the NI 9191 chassis (to the ground connection on the module housing), while the second end of the shield was disconnected,
- both ends of the shield were connected properly (to the negative potential wire of the MFC transducer and to the ground connection on the module housing of the NI 9191 chassis).

During measurements the disturbances were generated on the transmission wire with agreement to the generated by the testing system signal with frequency from 150 kHz up to 20 MHz. The frequency of the disturbance was changed with the step of 100 kHz at every 3 seconds. What is more the signal of disturbance was modulated with the frequency 2 Hz. It could be noticed that the proposed measuring system was sensitive to the influence of conducted disturbances. Only by application of shielded wire as the transmission wire and proper connection of the shield on both ends of the wire, the influence of conducted disturbances can be eliminated. In case of improper connection of the shielding, the results of measurements are affected by the signal of disturbances. It is very important to take it into account during measurements carried-out using proposed system based on the Macro Fiber Composite sensors.

In the paper [18] the temperature fluctuation during the device operation was tested. Laboratory tests were conducted on laboratory stand equipped with universal mechanical testing machine (Instron Electropuls 10000) and thermal chamber. During the tests samples were subjected to cyclic extortion simulating the operation of the system in various environmental conditions by forcing changes in system operation temperature with constant conditions of its excitation. The terminals of the tested piezoelectric transducer were connected to an external resistor (placed outside of the thermal chamber) and an electric voltage drop on it was recorded. An experimental analysis of the impact of changes in the working temperature of the tested system on the efficiency of electricity generated by the used MFC piezoelectric transducer was done. The test was carried out by measuring the values of the voltage drop on the connected resistor during the temperature changes of the piezoelectric transducer in the range from -30 to +70 Celsius degrees. The useful working range given by the manufacturer is from -35 to +85 Celsius degrees [2]. The tests were carried out with a temperature gradation every 10 °C. The piezoelectric transducer was heated at a given temperature for 10 minutes and then subjected to 500 load cycles by harmonically variable force with a frequency of 1.5 Hz. The laboratory stand was controlled using an extensometer, ensuring a constant deformation value of the transducer at the maximum level of 2000 ppm. A significant effect of temperature change on the voltage drop value on the resistor connected to the tested MFC piezoelectric transducer was observed. There was a decrease in the voltage peak value by more than 50% along with the temperature increase in the analysed range from -30 to +70 degrees Celsius. Such an important change translates directly into the efficiency of the energy recovery system from mechanical vibrations exposed to variable temperature operating conditions. One of the reasons for the decrease in the efficiency of the system as the temperature rises may be the change in the dielectric constant value of the material with piezoelectric properties, which is applied to the tested composite transducer. Conducted tests in laboratory conditions proved that temperature changes could significantly affect the efficiency of the energy harvesting system based on Macro Fiber Composite applications. By providing a constant value of deformation of the tested sample by using an extensometer to control movement of the machine jaws and providing a constant frequency, identical excitation parameters of the tested sample were ensured during all research tests. Only temperature changes could have influence on the obtained results.

4. Research on the impact of a mechanical subsystem damage on a system operation

To determine the impact on the measured values of damage of the mechanical subsystem on which the MFC-type piezoelectric transducer was glued, a series of laboratory tests was carried out on the created laboratory stand. The laboratory stand, shown in figure 1, consisted of the following elements:

- Tabor WW5064 25MHz 4-Ch Arbitrary Waveform Generator,
- High voltage amplifier HVA 1500/50,
- NI-cDAQ-91911-Slot, Ethernet and 802.11 Wi-Fi CompactDAQ Chassis with the measuring card NI-9215,
- Personal computer with LabVIEW software.

Composite samples made of carbon fiber in an epoxy resin matrix with dimensions of 200x40 mm and a thickness of 1 mm were tested. During the tests, the samples were clamped in the jaws of the vice, and two MFC M8514P1 type piezoelectric transducers were glued on their surface. One MFC transducer was used to excite the vibration of the tested sample and during the tests it was powered by a harmonically variable voltage signal of a given frequency and amplitude, generated by the W5064 generator and amplified by the HVA 1500/50 high voltage amplifier. The second of the MFC transducers was used as a sensor, and the value of the electric voltage generated by this transducer as a result of its deformation was measured using the NI-9215 measurement card and recorded using the NI-cDAQ-9191 measurement module in LabVIEW software.

Figure 1. Laboratory measuring stand with the tested sample and MFC-type transducers mounted on its surface.

Three samples were tested. The first of them had no damage (sample number 1 in figure 2), while the other two were damaged by making incisions in them. In the case of the first of the damaged samples, the incisions were irregular (sample number 2 in figure 2), while the second damaged sample had damage in the form of regularly spaced incisions of equal depth spaced evenly and running perpendicular to the sample axis (sample number 3). Damages were made to obtain changes in dynamic characteristics of the tested samples relative to the undamaged sample and to determine their impact on the values of the electric voltage generated by the MFC piezoelectric sensor.

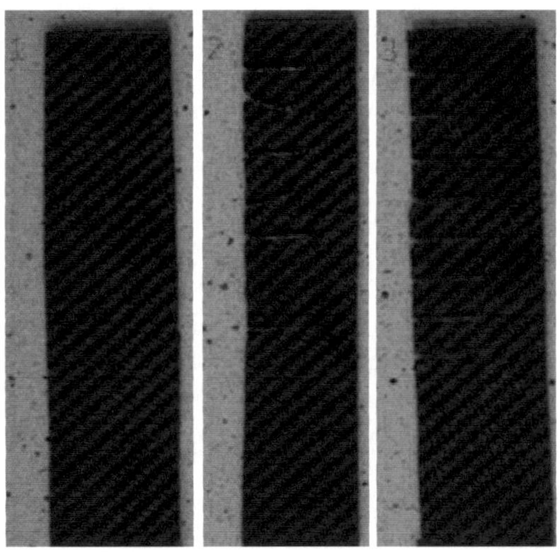

Figure 2. View of tested composite samples.

During the first stage of tests, the frequencies of natural vibrations of individual tested samples were determined by inducing them to vibrations using a signal in the form of noise. Signals generated by the MFC piezoelectric sensor were recorded and analyzed using Fast Fourier Transform. This way the first natural vibration frequency of the samples was determined. Then individual samples were excited to vibrate by supplying the piezoelectric transducer using harmonically variable signals with the frequencies equal to the determined first natural frequencies of the tested samples. At the same time, signals generated by the MFC piezoelectric sensor were recorded and the peak value generated by the sensor was determined. The results of the tests are listed in table 1.

Table 1. The peak voltage values generated by the tested samples depending on the excitation frequency.

	Sample 1 **No damage**	**Sample 2** **Irregular damage**	**Sample 3** **Regular damage**
Natural frequency	25.2 Hz	22 Hz	21.2 Hz
Excitation frequency 25.2 Hz	1.05 V	0.60 V	0.45 V
Excitation frequency 22 Hz	0.65 V	2,52 V	2.20 V
Excitation frequency 21.2 Hz	0.48 V	2.11 V	3.03 V

The obtained results indicate a significant impact of mechanical subsystem damages, on the surface of which MFC type piezoelectric sensors was glued, on the generated values of measurement signals. These differences result from changes in dynamic characteristics and the natural frequency of the tested samples as a result of their damage. The introduced damages caused changes of dynamic susceptibility of the samples and thus translated into higher values of deformation of the sensor glued on their surface. Higher vibration amplitude of individual samples in the range of their resonance frequency caused higher voltage peaks generated by the piezoelectric sensor. There is a visible increase in the value of voltage generated by the MFC sensor in the case of damaged samples relative to the sample without damage, resulting from their greater susceptibility and higher vibration amplitude.

5. Conclusions

The literature review, tests and measurements proved that Macro Fiber Composite piezoelectric transducers can be successfully used as sensors. They can easily be glued on the surface of tested elements and structures, as well as laminated in composite elements. What's more, they can be easily protected against weather conditions. However, as it was shown during the research on the impact of electromagnetic disturbances and changes in operating temperature, it should be remembered that the values of the electric voltage generated by the proposed sensors may be disturbed. Therefore, these changes should be considered during the design and construction process of mechatronic systems in which piezoelectric films are used. Omission of these phenomena can lead to incorrect measurements and incorrect operation of the system.

The tests carried out on composite samples made of carbon fiber in an epoxy resin matrix also indicate a high sensitivity of the measuring system based on the use of a piezoelectric sensor in the form of a MFC film glued on the surface of a vibrating mechanical subsystem on its damages and thus changes in dynamic characteristics. Any mechanical damage causes a change in the frequency and amplitude of vibration of the tested system and thus directly translates into the values of voltage generated by the piezoelectric sensor glued on the surface of the tested element. Therefore, the conducted research indicates that this type of sensors can be successfully used in systems for monitoring the technical condition of various technical means, including primarily elements made of composite materials. Such a solution was proposed in the case of testing the technical condition of composite panels used as the sheathing of boxes of modernized freight cars [1,8]. However, when measuring the vibration parameters of road infrastructure elements to detect the passage and collision of vehicles in vehicular traffic any damage and changes in the dynamic characteristics of the elements on which piezoelectric measuring sensors will be glued can lead to disturbances in the operation of the traffic monitoring system. Changes in the dynamic characteristics of the element on which the measurement is made can lead to a significant change in the signal generated by the MFC piezoelectric sensor and thus be misinterpreted, for example, as a vehicle collision. This phenomenon is extremely important in relation to the reliability and effectiveness of the proposed mechatronic road traffic supervision system and should be considered when selecting the elements on which the measuring sensors will be mounted and potential changes in their dynamic characteristics should be anticipated. These changes may be the result of exposure of system components, among others, to changing weather conditions, corrosive wear, mechanical damage and many other factors that should be foreseen and, if possible, reduce their impact on the designed measuring system. To this end, further research is required both in laboratory conditions and in real operating conditions to which the elements of the designed mechatronic measuring system based on the use of sensors in the form of piezoelectric films are subjected.

6. References

[1] Płaczek M, Buchacz A and Wróbel A 2015 Use of piezoelectric foils as tools for structural health monitoring of freight cars during exploitation *Eksploatacja i Niezawodnosc – Maintenance and Reliability* **17(3)** 443–449
[2] https://www.smart-material.com/index.html (Accessed 01.09.2019)
[3] Daue T P, Kunzmann J and Schonecker A 2008 Energy Harvesting System Using Piezoelectric Macro Fiber Composites *Smart Material Corporation*, Sarasota, FL, USA, 2008
[4] Voon-Kean W, Jee-Hou H and Hui-Ken S 2016 On accumulation of water drop-lets in piezoelectric energy harvesting *J. Intell. Mater. Syst. Struct.* **28** 521–530
[5] Gwiazda A, Herbuś K, Kost G and Ociepka P 2015 Designing mechatronics equipment based on the example of the Stewart platform *Solid State Phenomena* **220/221** 419-422
[6] Buchacz A and Gałęziowski D 2016 Designing of discrete mechatronic vibrating systems with negative value parameters *Mechanical Systems and Signal Processing* **78** 134-142
[7] Płaczek M, Wróbel A and Buchacz A 2017 Structural Tests of Freight Wagons on the Basis of Signals Generated by Piezoelectric Macro Fiber Composites *Acta Mech. et Automatica* **11(3)** 210-216
[8] Buchacz A, Baier A, Świder A et al. 2016 *Analytical and experimental tests and determination of characteristics of components working as assemblies of innovative structures of repaired freight cars* (Gliwice: Silesian University of Technology Publishing House)
[9] Sekuła K, Kołakowski P, Świercz A 2010 System for the monitoring of loads and technical state of truss bridges in the railway *Condition Monitoring and Evaluation of Technical Construction and Its Vitality*, Seminar MONIT, Warsaw, 19.11.2010: 1-4
[10] Stypuła K 2009 Selected problems of surface building protection against vibrations generated by underground communication *Górnictwo i Geoinżynieria* **3(1)** 351-362
[11] Chłus K, Krasoń W 2011 Simulation of the effect of static and dynamic loads on the strength of the railway platforms *Article Proprietary X Eng. Forum ProCAx*, Sosnowiec/Siewierz. 6-9.10.2011
[12] Grebowski K, Zielińska M 2015 Modelling of dynamic interactions of Pendolino train type on structures of historic railway bridges in Poland *Przegląd Budowlany* **1** 27-32
[13] Herwig A, Bruhwiler E 2011 In-situ dynamic behaviour of a railway bridge girder under fatigue causing traffic loading *Proc. of the 11th International Conference on Applications of Statistics and Probability in Civil Engineering, ICASP11*, Zurich, 1-4 August 2011 389-395.
[14] Oleszak P, Cieśla J and Szaniec W 2013 Study of the effects of side impacts on railway viaduct lying on the arc *Budownictwo i Architektura* **12(2)** 47-54
[15] Płaczek M 2016 Conception of the system for traffic measurements based on piezoelectric foils *IOP Conf. Ser.: Mater. Sci. Eng.* **145** 042025
[16] Płaczek M 2011 Indication of the suitable model of a mechatronic system as an introduction to the synthesis task *J. Achiev. Mater. Manuf. Eng.* **49/2** 338-349
[17] Płaczek M and Maćkowski M 2018 Electromagnetic compatibility tests of a measuring system based on MFC piezoelectric transducers *IOP Conf. Series: Mat. Science and Eng.* **400** 022046
[18] Płaczek M and Kokot G 2019 Modelling and Laboratory Tests of the Temperature Influence on the Efficiency of the Energy Harvesting System Based on MFC Piezoelectric Transducers *Sensors* **19(7)** 1558
[19] Lin X, Chen H, Ma Y et al. 2017 Investigation of temperature sensitivity of actuation performance for piezoelectric fiber composites *Ceram. Int.* **43** 10590–10594
[20] Williams R B, Inman D J and Wilkie W K 2004 Temperature-Dependent coefficients of thermal expansion for Macro Fiber Composite actuators *J. Therm. Stresses* **27** 903–915.

Aspects on the study of nonlinear vibration phenomenon in friction welding with rotating active element

C Dragomirescu[1] and V Iliescu[2]

[1] Professor, Polytechnic University, Bucharest, Romania
[2] Professor, University for Agronomical Sciences and Veterinary Medicine, Bucharest, Romania

victor.iliescu@ymail.com

Abstract. The friction welding with rotating active element is a complex process including heating of the materials using friction and plastic deformation generated by a tool that has both rotation and translation motions. The paper is analysing the nonlinear vibrations that occurs during the friction welding with rotating active element process, mainly due to the motion of the rotating element. Using the dynamic systems theory, the nonlinear equation of the motion is deduced and the motion characteristics are presented using specific qualitative methods. The results are contributing to a more accurate understanding of the phenomena, in this case the nonlinear vibrations, generated in the friction welding with rotating active element process, and to better performances of welding process.

1. Introduction

The idea of jointing two materials using friction emerged at the end of the 19-th century and is described by J. H. Bevingtonin a patent, and used it to connect wires, edges and sleeves. The interest for this method continued in the XX-th century, the researches of Schaefer (1971), Meyer (1978) and Kes (1989) analysing the effects of the friction welding. However, the development of the technology was not at the standard required by the industry, at least in the first half of the century [1]. The soviet researchers also studied the method and developed an industrial technology for friction welding (a patent of A. I. Chudikov in 1956), later upgraded in SUA, Japan, UK and Germany. The result was a patent obtained by Caterpillar Tractor Co. in 1966, for “welding by inertial friction” [1].

Friction welding is now used by decades in industry for jointing and was initially known as “welding by rotation” and later, considering the linear welding, as “motion cinematic”, a pleonastic expression. Even nowadays, a lot of persons understand the “friction welding” through these two processes.

At present, other varieties of welding process are in place: orbital welding, circular welding for rectangular elements, rotational vibrations welding for round shaped elements (having high performance requirements) etc. The innovations in precision friction welding were focused on the operating technology for highly dynamic friction welding, precision measuring technology, friction welding technology in manufacturing, protection devices used in friction welding, control process in friction welding, real time communication of data in friction welding, increasing the productivity in friction welding by shortening the welding time etc. [2]. These areas of interest were covered by theoretical studies and experiments, both aiming to clarify and improve this technology. We are pointing out, only as examples, some researchers: De Backer & Bolmsjö [3], Gheorghiu & Bejinariu

[4], Langhorst, Pöhler, Zaeh & Lotz [5], Lohe, Lotz, Cannon & Kouvaritakis [6], Schmidt & Hattel [7], Shinoda, Miyahara, Ogawa & Endo [8], Wang & Wen [9], Zäh & Voellner [10] etc.

Friction welding with rotating active element is a technology invented in 1991 at Cambridge, at the Institute for Welding Technologies [6]. This technology is allowing jointing two elements without melting the material, without pressing one to the other and without built-up material. The friction welding with rotating active element is a fast and inexpensive welding technology.

Friction welding with rotating active element is mainly used for the welding of aluminium and its alloys, due to the fact that the obtained joints are solid and with good plasticity. The welded elements are mainly flat and long (plates) but may be adapted for other thin profiles.

Friction welding with rotating active element is the result of combined friction heating and plastic deformation, using tool having both rotation and translation motions. The maximum temperature reached is 80% of the melting temperature. The diagram of the process is shown in figure 1 and the basic description of the process of friction welding with rotating active is the following. The active rotating element is having two parts: a massive part used to fix the tool in the machine and a slenderer, pointed part or "the pin" – the active part of the tool. After the proper positioning and fastening of the two work pieces to be welded, the active element is starting to rotate and to move toward the two pieces. The pin is penetrating the surface of the work pieces up to the massive part of the tool called "shoulder". Due to the friction with the shoulder, the two work pieces are heated, obtaining a plastic area in the two work pieces to be welded. At this moment, the active rotating element is starting the translation motion on the welding line. As the active element moves, the heated material (by conduction) in the front of the pin is displaced in the gap behind the pin, generated by the translation of the pin. The shoulder of the active element is pressing down the welding, resulting a flat even weldseam. Any oxide film on the surface of the two pieces is dispersed due to the intense deformation. The process of local deformation is similar to a continuous extrusion process along the joint.

The section of the point may be a circle or a polygon, mainly a triangle or a square. The point may have a constant or variable section (i.e. a cylinder or a truncated cone) [4].

In the friction welding with rotating active element, the shape of the point, the rotation speed of the active element, the translation speed of the active element, the pressure applied by the active element, the characteristics of the welded materials are generating different configurations and mechanical characteristics of the bead of weld obtained.

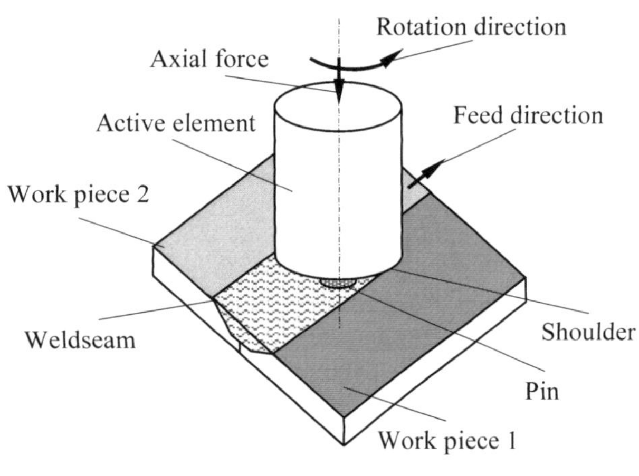

Figure 1. Schematic diagrame – Friction welding with rotating active element

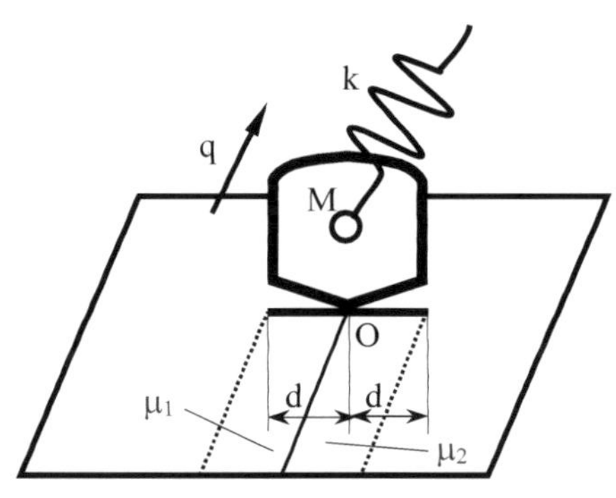

Figure 2. The mechanical model used for the study

2. The model used for the study

The paper is developing an analytical study of the friction process generated by the translation motion of the active element. The approach is based on the nonlinear vibrations phenomena and is considering only the qualitative aspects at this time. Accordingly, we adopted the model in figure 2, where M is the mass of the active element, k is the elastic coefficient of the active element, $2d$ is the width of the weldseam. The positions of the pin and the position of the elastic element are measured from O point.

The friction coefficients, μ_1 and μ_2 (considering the values in the plastic state of the workpieces) are not having the same value, so the friction force $\overline{F}_f$ generated between the tip and the work pieces is having different values and the differential motion equation may be written as follows [11], [12]:

$$M\ddot{q} = \begin{cases} -F_{f1}, & q \le d; \dot{q} > u \\ -F_{f2} - k(q-d), & q > d; \dot{q} > u \\ 0, & q \le d + \dfrac{F_{f1}}{k}; \dot{q} = u \\ F_{f1}, & q \le d; \dot{q} < u \\ F_{f2} - k(q-d), & q > d; \dot{q} < u \end{cases}, \tag{1}$$

where u is the speed of the pin.

Integrating the motion equation (1) we obtain the following solution:

$$q = \begin{cases} -\dfrac{1}{2}R_1 t^2 + C_1 t + C_2, & q \le d; \dot{q} > u \\ C_3 \sin(\omega t) + C_4 \cos(\omega t) + d - \dfrac{F_{f2}}{k}, & q > d; \dot{q} > u \\ ut + C_5, & q \le d + \dfrac{F_{f1}}{k}; \dot{q} = u \\ \dfrac{1}{2}R_1 t^2 + C_6 t + C_7, & q \le d; \dot{q} < u \\ C_8 \sin(\omega t) + C_9 \cos(\omega t) + d + \dfrac{F_{f2}}{k}, & q > d; \dot{q} < u \end{cases}, \tag{2}$$

where $R_1 = \dfrac{F_{f1}}{M}$, $\omega^2 = \dfrac{k}{M}$ and $C_1 - C_9$ are integration constants obtainable from the initial conditions:

$$q = q_0 < d; \ f_0 = \dot{q}_0 > u. \tag{3}$$

If we are studying the phase plane trajectories from a qualitative point of view, we may use the variable $f = \dot{q}$ in order to write the equations of the phase trajectories:

$$f^2 = \begin{cases} -2R_1 q + C_1^2 + 2R_1 C_2, & f > u; q \le d \\ \omega^2 \left\{ C_3^2 + C_4^2 - \left[q - \left(d - \dfrac{F_{f2}}{k} \right) \right]^2 \right\}, & f > u; q < a \\ u^2, & f > u; q < d + \dfrac{F_{f1}}{k} \\ 2R_1 q + C_6^2 - 2R_1 C_7, & f < u; q \le d \\ \omega^2 \left\{ C_8^2 + C_9^2 - \left[q - \left(d + \dfrac{F_{f2}}{k} \right) \right]^2 \right\}, & f < u; q > a \end{cases}. \tag{4}$$

3. Results interpretation

The numeric simulation was performed using MATLAB, obtaining sets of graphics, representing the evolution in time of the solutions (2) and the associated phase plane trajectories (mainly parabola – see Table 1) using equation (4) and considering the condition (5). In figure 3 and figure 4 we are showing an example of a set of graphics, calculated considering k = 1 N/m, F_{f1} = 5 kN, u = 3.927 m/s, M = 250 kg and d = 2 mm [1], [2], [4].

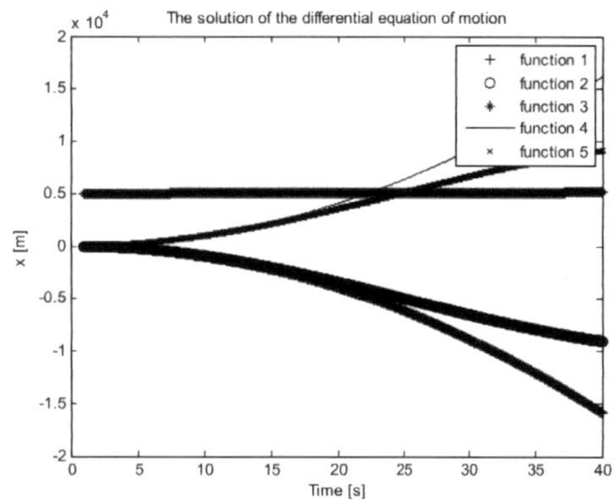

Figure 3. The time variation of the solutions (2)

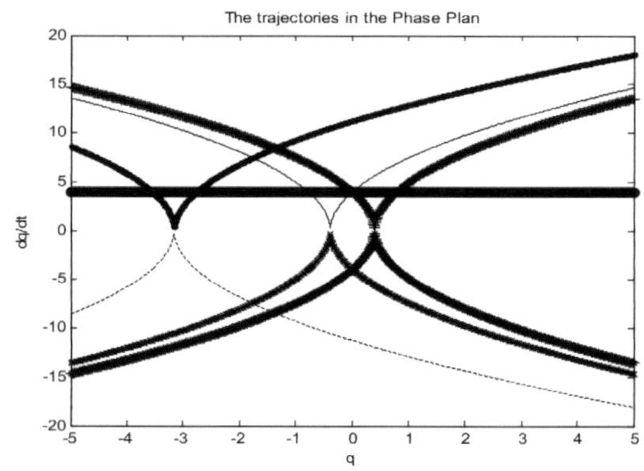

Figure 4. Phase plane trajectories

Because, for the moment, we want only to point out the conditions for a stable vibrational motion, we adjusted, as scale, some representative phase plane trajectories and overlapped them in order to obtain an overall, *general qualitative* image of the phenomenon (see figure 5).

$$k < \frac{F_{f2}^2}{Mf_0^2}, \tag{5}$$

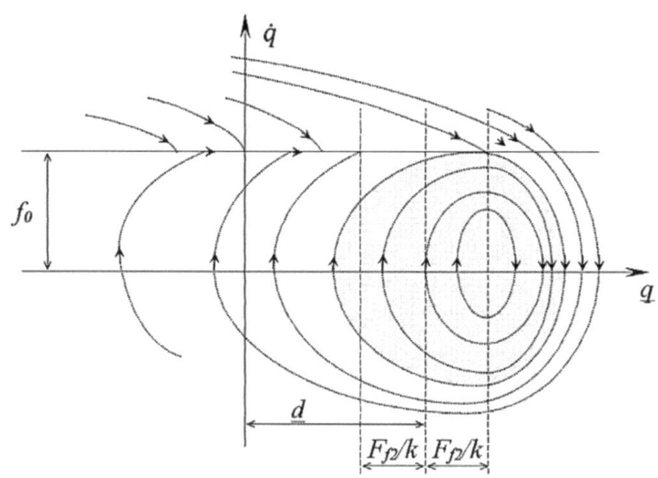

Figure 5. Phase plane $(q,\dot{q})$ trajectories – general qualitative

Table 1. The results interpretations [13-15]

No	Case	Description of the motion
1.	$f > f_0; q \le d$	Parabola intersects the straight line $f = f_0$ for $q < d$ and the representative point is moving on this line up to point $A\left(d + \frac{F_{f2}}{k}, f_0\right)$.

<table>
<tr><th>No</th><th>Case</th><th>Description of the motion</th></tr>
<tr><td>2.</td><td>$f < f_0; q > d$</td><td>For $k > \frac{F_{f2}^2}{Mf_0^2}$, the representative point is moving on an ellipse, from A to $B\left(d, -\sqrt{f_0^2 - \frac{F_{f2}^2}{Mk}}\right)$, situated at the intersection with the straight line $q = d$.</td></tr>
<tr><td>3.</td><td>$f < f_0; q \le d$</td><td>The motion is on a parabola up to point $C\left(d, \sqrt{f_0^2 - \frac{F_{f2}^2}{Mk}}\right)$, for $k > \frac{F_{f2}^2}{Mf_0^2}$.</td></tr>
<tr><td>4.</td><td>$f < f_0; q > d$</td><td>The representative point is moving on the same ellipse as in case 2, up to point A, then the motion is repeating periodically.</td></tr>
<tr><td colspan="3">Obs.: If the condition (5) is not fulfilled, the representative point is moving on the same ellipse, not intersecting the straight line $q = d$.</td></tr>
<tr><td>5.</td><td>$f > f_0; q \le d$</td><td>Parabola intersects the line $q = d$ in $D(d, f_1)$, for $f_1 > f$.</td></tr>
<tr><td>6.</td><td>$f > f_0; q > d$</td><td>Out of the initial conditions $q(0) = d; f(0) = f_1$, one may calculate the integration constants $C_3 = \frac{f_1}{\omega}$ and $C_4 = \frac{F_{f2}}{k}$. The representative point is moving on an ellipse from D until the phase trajectory is intersecting the straight line $f = f_0$, for $x < d + \frac{F_{f2}}{k}$, then the representative point is moving on this straight line up to the point A. Afterwards the motion is similar to one in the case 1.</td></tr>
<tr><td>7.</td><td>$f < f_0; q > d$</td><td>Out of the initial conditions $q(0) = q_1; f(0) = f_0$, one may calculate the integration constants $C_8 = \frac{f_0}{\omega}$ and $C_9 = q_1 - \left(d + \frac{F_{f2}}{k}\right)$. If the phase trajectory is intersecting the straight line $f = f_0$ in the point $E\left(q_1 = d - \frac{F_{f2}}{k} + \sqrt{\frac{F_{f2}^2 + M(f_1^2 - f_0^2)}{k}}, f_0\right)$, for $q_1 \ge d + \frac{F_{f2}}{k}$, the representative point is moving also on an ellipse, different from the one above, due to the different initial conditions.</td></tr>
<tr><td>8.</td><td>$f < f_0; q > d$</td><td>The motion on the ellipse in case 7 is possible until the phase trajectory is intersecting the straight line $q = d$ in the point $F\left(d, \sqrt{f_0^2 - \frac{F_{f2}^2}{Mk} + \left[\sqrt{\frac{F_{f2}^2}{Mk} + f_1^2 - f_0^2} - 2\frac{F_{f2}}{\sqrt{Mk}}\right]^2}\right)$.</td></tr>
<tr><td>9.</td><td>$f < f_0; q \le d$</td><td>The motion of the representative point is on a the parabola up to the intersection with the straight line $f = f_0$, then the motion is on this straight line up to the point A, then the motion is repeating like in the case 1.</td></tr>
<tr><td colspan="3">Obs.: For the initial conditions $q = q_0 < d; f = f_0 < u$, the representative point is moving first on the parabola $f < f_0; q \le d$ and the study is reduced to the case 1. When the initial conditions are generating a dot in the marked domain in Figure 5, the motion of the representative point is going to be periodically from the beginning.</td></tr>
</table>

IOP Conf. Series: Materials Science and Engineering **682** (2019) 012011 doi:10.1088/1757-899X/682/1/012011

4. Conclusions

During the welding process, self-sustained vibrations are generated by the dry friction. These vibrations are nonlinear due to the friction force, as shown above.

The results interpretations (Table 1) are allowing us pointing out the characteristics of the solutions describing periodical motions, considering the weldseam width and the feed rate. This means that the possible pin trajectories may be identified, helping to better managing the technological process. Also, the more stable vibrations may be identified.

The less dangerous vibrations are the ones with the representative point moving on an ellipse or on a part of an ellipse: cases 2, 4, 7 and 8 from Table 1, because the motion is more predictable and a slight adjustment of the welding parameters may avoid any problem.

A possible problem may occur in the cases when the representative point is moving on a parabola or on a parabola and a straight line: cases 1, 3, 5, 6, 9, because the motion is less controllable and very small changes of the parameters may totally change the motion type. A special mention for the case 6, where the motion of the representative point is starting on an ellipse (meaning more stable motion) and then changing to the case 1.

The *general qualitative* results obtained are demonstrating that the study must go deeper in the *quantitative* aspects of the problem, in order to see which welding parameters or which constructive parameters may be better adjusted for a more stable motion.

5. References

[1] Lotz M J 2012 *Erhöhung der Fertigungsgenauigkeit beim Schwungrad-Reibschweißen durch modellbasierte Regelungsverfahren, Dissertation* (München: Fakultät für Maschinenwesen der Technischen Universität München)

[2] Raab U 2016 *Untersuchungen zum Orbital-Reibschweißen als alternatives Herstellungsverfahren von Titan-Blisks im Vergleich mit dem Linear-Reibschweißen,* Dissertation, Fakultät für Natur - und Materialwissenschaften (Technischen Universität Clausthal)

[3] De Backer J and Bolmsjö G 2014 Deflection model for robotic friction stir welding *Ind. Robot* **41(4)** 365-372

[4] Gheorghiu D A and Bejinariu C 2013 *Metalurgia sudării* (Iaşi: Editura PIM)

[5] Langhorst M, Pöhler A, Zaeh M F and Lotz M 2009 The Structural Behaviour of an Inertia Friction Welding Machine *NAFEMS World Congress (Crete, Greece, 16-19 June 2009)*

[6] Lohe J, Lotz M, Cannon M and Kouvaritakis B 2013 Application of Optimal Control Algorithm to Inertia Friction Welding Process *IEEE Trans. Control Syst. Technol.* **21(3)** 891–898

[7] Schmidt H and Hattel J 2005 A local model for the thermomechanical conditions in friction stir welding *Model. Simul. Mater. Sci. Eng.* **13(1)** 77–93

[8] Shinoda T, Miyahara K, Ogawa M and Endo S 2001 Friction welding of aluminium and plain low carbon steel *Weld. Int.* **15(6)** 438–445

[9] Wang K K and Wen L 1974 Flywheel friction welding research *Weld. J.* **53(6)** 233–241

[10] Zäh M and Voellner G 2010 Three-dimensional friction stir welding using a high payload industrial robot *Prod. Eng.* **4(2-3)** 127–133

[11] Deciu E and Dragomirescu C 2001 *Maschinendynamik* (Bucharest: PRINTECH)

[12] Fischer U and Stephan W 1984 *Mechanische Schwingungen* (Leipzig: VEB Fachbuchverlag)

[13] Deciu E, Bugaru M and Dragomirescu C 2002 *Vibraţii neliniare cu aplicaţii în ingineria mecanică* (Bucharest: Editura Academiei Romane)

[14] Dincă F and Teodosiu C 1969 *Vibraţii neliniare şi aleatoare* (Bucharest: Editura Academiei R.S.R.)

[15] Voinea R and Stroe I 2000 *Introducere în teoria sistemelor dinamice* (Bucharest: Editura Academiei Române)

IOP Conf. Series: Materials Science and Engineering **682** (2019) 012012 doi:10.1088/1757-899X/682/1/012012

Use of the Ishikawa diagram in the investigation of some industrial processes

C Botezatu[1], I Condrea[1], B Oroian[1], A Hrițuc[2], M Ețcu[1] and L Slătineanu[3]

[1] Ph. D. Student, Technical University of Iași, Romania
[2] M.Sc. Student, Technical University of Iași, Romania
[3] Professor, Technical University of Iași, Romania

hrituc.adelina3295@yahoo.com

Abstract. The use of the Ishikawa diagram was proposed by Japanese professor Kaoru Ishikawa in the 1960s. The diagram must highlight the possible causes of a certain effect. In the research presented in this paper, the analysis method based on the design of the Ishikawa diagram was used to identify the factors able to ensure the adequate development of some investigations concerning certain industrial processes. It was found that various criteria could be used to evaluate the effect. Versions of Ishikawa diagrams were proposed in the cases of chemical engraving process, cast iron milling process, 3D printing process and the reliability of computer subsystems, respectively. A comparison of the Ishikawa diagram method with the systemic analysis method highlighted some particular aspects that could be considered when the problem of investigating the influence of certain process input factors on the process output parameters is formulated.

1. Introduction

The main groups of theoretical research methods are the analysis methods, synthesis method, modeling and/or simulation methods, optimizing methods. One of the theoretical analysis methods is the analysis through the cause-effect method. This method was proposed by the Japanese professor Kaoru Ishikawa, in the 1960s, when he worked at Tokyo University. Sometimes, the application of the method is finalized by the elaboration of the so-called Ishikawa diagram or diagram in the fishbone or herringbone diagram.

Practically, the Ishikawa diagram is a graphical representation of the connection between a result and the factors able to exert influence on the result. In this way, for example, the diagram offers a general image concerning the causes that generate a certain effect. The main stages in the elaboration of the Ishikawa diagram are the following: *1.* Problem definition; *2.* Elaboration of the graphical representation; *3.* Analysis of the information offered by the diagram and that highlights the main factors or the factors whose values could be changed; *4.* Developing a plan to act by considering the remarks from the previous stage.

As benefits of using the Ishikawa diagram, the following could be mentioned: a better understanding of a process or equipment, initiation of an innovation process, facilitating the learning process and the ideas change, better manage of the factors able to generate less convenient effects, eventual establishing of the necessity of technical norms elaboration etc.

Over the years, the Ishikawa diagram was analyzed or used in solving many scientific research problems.

IOP Conf. Series: Materials Science and Engineering **682** (2019) 012012 doi:10.1088/1757-899X/682/1/012012

Thus, Ilie and Ciocoiu took into consideration the risk of an event with multiple causes in the elaboration of the fishbone diagram [1]. They considered that the inclusion of the risk score for each category of causes could extend the Ishikawa diagram applicability.

Romo et al. showed how the Ishikawa diagram could be used to identify the possible causes of a problem [2]. They highlighted the importance of using subsequently the question "why does this happen" in problem-solving. The importance of considering the correct evaluation of the probability that corresponds to various causes and their impact was mentioned.

Magdum and Gore investigated a manufacturing process to identify the root causes of parts rejection and found that various factors generate a certain ovality of the part [3]. The analysis facilitated the elaboration of an Ishikawa diagram, used then as a guideline for solving the problem. They appreciated that other methods could be applied to prioritize the causes that generate the part rejection and to apply a certain solution.

Luca developed research concerning the errors that appear in the evaluation of the precision of the parts. He took into consideration the influence of various factors on the dimensional accuracy, on the shape and position accuracy and the surface roughness, respectively [4]. On the other hand, five groups of factors were identified as able to affect the parts precision evaluation: man, method, machines, materials, and environment. The author proposed the use of an improved model of Ishikawa diagram, obtained as a combination of three skeletons that correspond to considered ways of parts precision evaluation.

Luo et al. highlighted the possibility of using the fishbone diagram and risk analysis method in the safety evaluation of a natural spherical gas tank [5]. They proposed the use of a modified classical risk matrix in a combination with an improved version of the fishbone diagram. As the main results, they mentioned the elaboration of specific measures to eliminate or to diminish the risk of accidents.

The objective of the research presented in this paper was to highlight the possibilities of using the Ishikawa diagram in initiating some doctoral research activities that correspond to certain processes or equipment from the industrial engineering field. The connection between these processes is that having some input factors, starting from a raw part, and needed some output factors, generally shaping the finished part, using the Ishikawa diagram, someone can see some common lines to guide within the direction of solving the problems occurring during the technological processes.

2. Use of Ishikawa diagram in defining the conditions for experimental research of chemical engraving process

The chemical machining can be defined as a non-traditional machining method based on the developing of the chemical reaction between a chemically active liquid and the workpiece material, in the workpiece zones where a process of material removal is necessary. One of the proper applications of the chemical machining method is the chemical engraving [6]. The last decades showed an extension of using the chemical engraving to perform inscriptions of drawings on the metallic or nonmetallic parts. As a chemical machining method, the chemical engraving needs to know and adequately use a set of material conditions and process characteristics. With such an aim in view and taking into consideration the facilities offered by the Ishikawa diagram and the research activity intended to be developed on the chemical engraving process, the decision to elaborate such a graphical representation was formulated.

As engraving process output results, the material removal rate, the surface roughness, the thickness of the layer affected by the machining process and the change of the chemical composition and physical properties of the chemically active liquid were considered.

The information found in the specialty literature and the analysis of the factors that are able to affect the process results (the values of the process input factors could be included in the following main groups: specimen characteristics, chemical engraving scheme, characteristics of the chemical engraving equipment, human factor involved in the investigated process) were considered. The deeper analysis of the chemical engraving process led to the identification of proper factors included in the above-mentioned groups and their consideration within an Ishikawa diagram (figure 1).

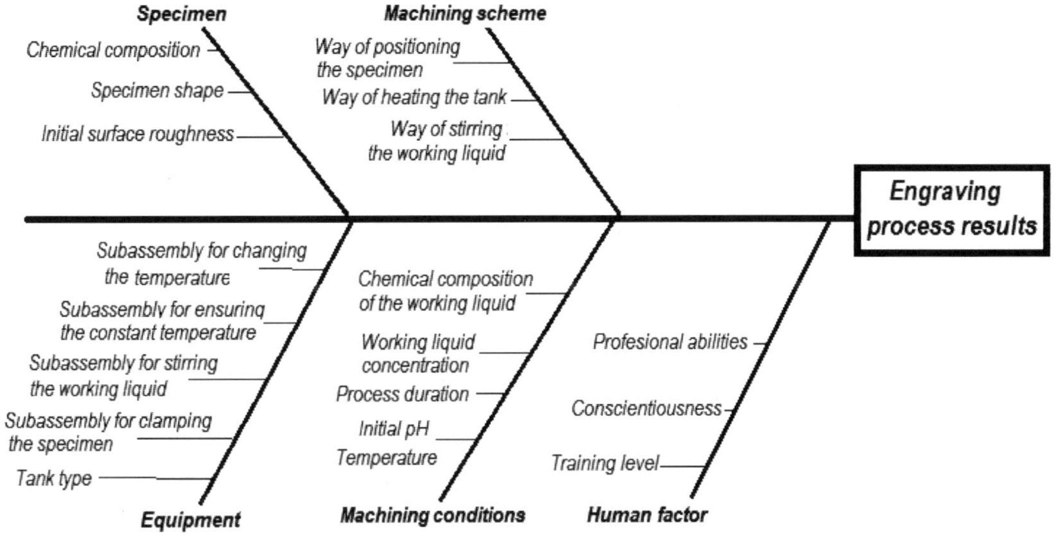

Figure 1. Ishikawa diagram elaborated to highlight the factors able to affect the results of the chemical engraving process.

3. Use of Ishikawa diagram to identify the factors able to affect the reliability of the computer

At present, the computers can not only perform complicated calculations, but they can be embedded in systems capable of solving complex problems, with or without human help. Current computers include many components whose operation can significantly affect the good functioning of the whole ensemble in which these components are incorporated. There are many known incidents caused by the malfunctioning of some components of computers. To highlight the factors able to exert influence on the reliability of the computers, the method of the Ishikawa diagram was used (figure 2). The main groups of factors affecting the good functioning of the computers could be the following: *a)* Power supply; *b)* Component reliability; *c)* Environment conditions; *d)* Human factor; *e)* Operating conditions. The proper factors that act on the computer's reliability and that determine their good functionality were emphasized in the Ishikawa diagram presented in figure 2.

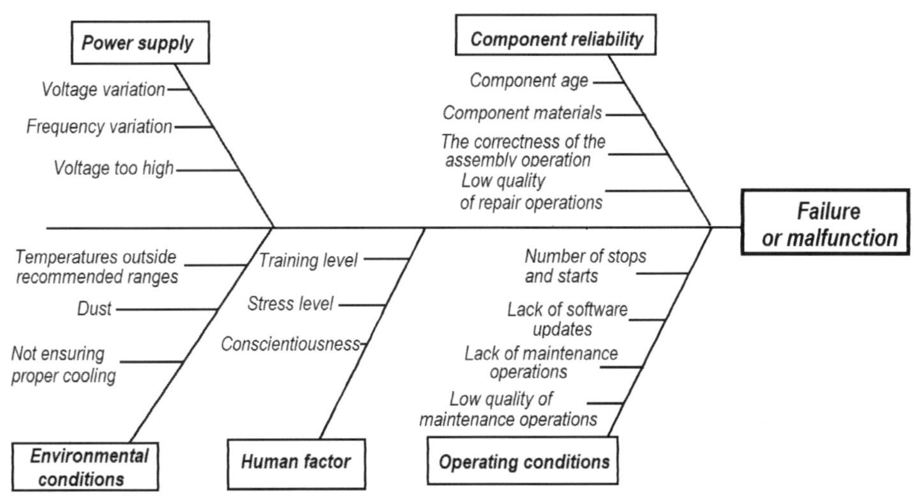

Figure 2. Ishikawa diagram that illustrates the factors able to determine the failure or malfunction of the computer system.

4. Use of Ishikawa diagram in evaluating the results of the cast iron end milling process

The milling is a machining method based on the use of a rotating cutting tool that has many cutting edges. The feed movement is performed along a direction perpendicular to the cutting tool rotation axis. The milling is applied especially in obtaining flat surfaces, but it could be also applied to obtain cylindrical, conical, spiral or more complex profiled surfaces.

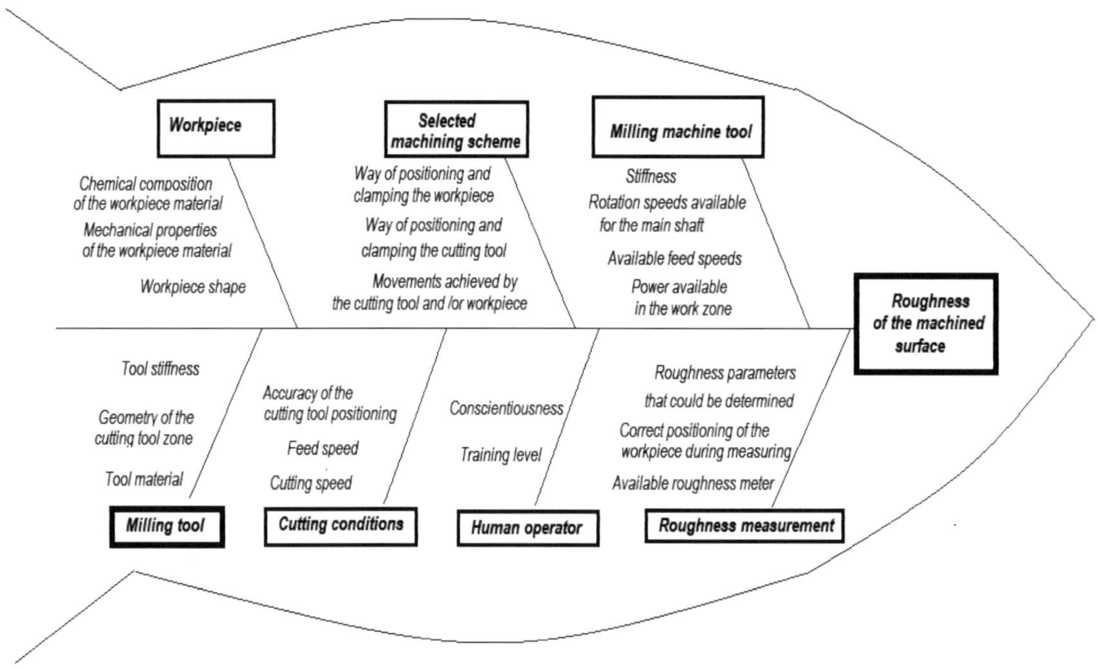

Figure 3. Ishikawa diagram that highlights the factors able to affect the roughness of the surfaces obtained by milling on cast iron workpieces.

The problem taken into consideration within a doctoral research program was to find solutions for improving the milling process of flat surfaces in the case of workpieces made of iron surfaces. As the main objective of the research, the improving of the surface roughness parameters was considered, aiming at finding the factors that could lead to the diminishing of the asperity's heights of the machined surface. With this aim in view, the Ishikawa diagram presented in figure 3 was elaborated.

5. Use of Ishikawa diagram in the evaluation of the 3D printing process

The 3D printing process seems to conquer more interest based on multiple utilization in the industry. The fused filament fabrication and selective laser sintering are only some of the well-known methods used for prototyping that uses the layer by layer deposition. The idea of obtaining parts of plastic materials with properties comparable with those specific to the metallic materials and made by 3D printing continues to be investigated and developed.

As far as 3D filament printing is concerned, complex shapes of the piece and no special tooling are considerable advantages to be preferred for the parts manufacturing. Also, the process is quite simple, consisting of designing the piece and transferring the 3D model to the software of the equipment, preparing and melting the material that is subsequently pushed through a nozzle, layer by layer, to gradually generate the desired part.

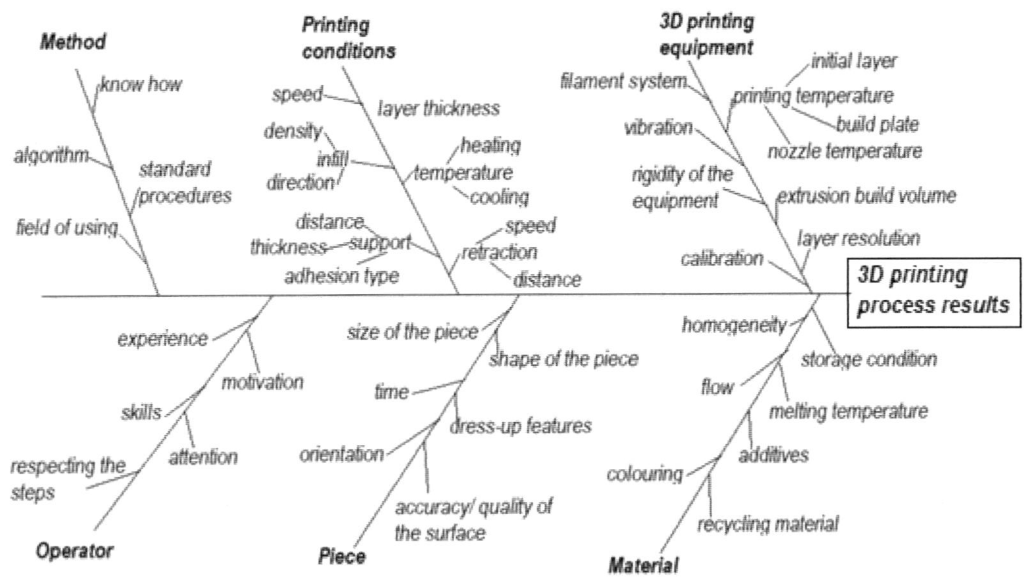

Figure 4. Ishikawa diagram that illustrates the factors able to affect the 3D printing process results

The process is influenced by some general factors, including equipment, the part to be obtained, printing conditions, the method with all the necessary steps to be followed, but also the operator. These factors were included in the Ishikawa diagram presented in figure 4, considering the highlighting of the factors able to affect the 3D printing process as a problem to be investigated and fulfilling thus the stage of the problem definition when using the Ishikawa diagram.

6. Comparison of the Ishikawa diagram method with the systemic analysis method

There were many preoccupations to apply and improve the use of the Ishikawa diagram method in solving a large set of technical and/or managerial problems and the fishbone type tool proved its utility in various situations.

The above-mentioned reasons led to the use of the Ishikawa diagram methods inclusively in solving problems from the industrial processes. The users were frequently interested in using adequate methods to identify the factors able to exert influence on the product and processes quality.

When using an industrial process, certain methods could be applied to investigate the influence exerted by distinct factors on the values of the process output parameters and the correlations between the two categories of factors. Such an investigation method could be a systemic analysis [7]. In principle, the systemic analysis considers the equipment or process investigated *as a system*, to which are assigned input factors, output parameters and possibly disturbing factors [8]. In the case of a process, such as the industrial processes previously addressed by using the Ishikawa diagram, *as input factors*, the groups of factors and the actual factors identified when the graphical representation corresponding to the Ishikawa diagram has been developed can be mentioned. As the output parameters of the investigated process, in the case of systemic analysis, several sizes of technological interest may be considered. In more complex representations, the actual connections between certain input factors and certain output parameters of the investigated process can be highlighted. The last two aspects may constitute the advantages of using systemic analysis comparing to the use of the Ishikawa diagram method.

To develop a comparison of the Ishikawa diagram method and a systemic analysis method, the following criteria could be taken into consideration:

- Method simplicity;
- Use of multiple output parameters;

IOP Conf. Series: Materials Science and Engineering **682** (2019) 012012 doi:10.1088/1757-899X/682/1/012012

- Inclusion of correlations among the input factors and output parameters;
- The necessity of using yet other methods (combination or association of the investigated analysis method with other methods);
- Suggestibility.

A weighting of the evaluation criteria could be made using the so-called matrix with double entries (or pairwise comparison matrix) (table 1) [8, 9], in which the results of comparing the identified criteria each other were included. The criteria were highlighted both along the first horizontal line and the first vertical column, but the comparisons were made firstly taking into consideration the vertical columns. Marks of 1 and 0 were inscribed when the criterion placed in the vertical column is appreciated as being more important than the criterion included in the horizontal line, and 0-1 when the situation was inverse. Marks of 0.5-0.5 were used when the two considered criteria were appreciated as of equal importance. In the penultimate horizontal line of the table, the sum of marks that correspond to each criterion mentioned in a certain vertical column was mentioned. In the last horizontal line of the table, the calculated values of the coefficients of relative importance were included. The coefficient of relative importance was determined as a ratio of the sum of marks to the number of comparisons applied to the criterion considered along the vertical column (there were such 4 comparisons).

Table 1. The weighting of evaluation criteria used to compare two distinct analysis methods.

	Simplicity	Multiple output parameters	Highlighting the correlations	Association with other methods	Suggestibility
Simplicity	X	1.0	0.5	0.5	0.5
Multiple output parameters	0	X	1.0	0	0
Highlighting the correlations	0.5	0	X	0	0
Association with other methods	0.5	1.0	1.0	X	0.5
Suggestibility	0.5	1.0	1.0	0.5	X
Sum of marks	1.5	2.0	3.0	1.0	1.0
Importance relative coefficient	0.37	0.5	0.75	0.25	0.25

In the second stage, comparisons of each analysis method using each of the comparison criteria were made. The results of these comparisons were presented in table 2.

Table 2. Comparison of two analysis methods.

Criterion / **Analysis method**	Simplicity	Multiple output parameters	Highlighting the correlations	Association with other methods	Suggestibility
Ishikawa diagram method	1	0	0	1	1
Systemic analysis	0	1	1	0	0

Taking into consideration the information included in tables 1 and 2, the coefficient of importance N_{vi} for each of the two methods could be calculated [9, 10]. The value of this coefficient is given by the relation:

$$N_{vi} = \sum_{1}^{4} K_{ij} K_j \tag{1}$$

where i is the number of compared methods, j – the comparison criterion, K_{ij} is the values of the coefficients of importance associated with each of the methods i and for each of the comparison criterion j.

In this way, the following values of the coefficients of importance are determined:

- In the case of the Ishikawa diagram method:

$$N_{v1} = 0.37 \cdot 1 + 0.5 \cdot 0 + 0.75 \cdot 0 + 0.25 \cdot 1 + 0.25 \cdot 1 = 0.87 \tag{2}$$

- In the case of the systemic analysis method:

$$N_{v2} = 0.37 \cdot 0 + 0.5 \cdot 1 + 0.75 \cdot 1 + 0.25 \cdot 0 + 0.25 \cdot 0 = 1.25. \tag{3}$$

The above relations show that the Ishikawa diagram has a coefficient of importance of 0.87, while the systemic analysis method seems to be of 1.25 importance coefficient.

On the determined values of the coefficient of importance, as a global aspect, the systemic analysis could be considered as more advantageous, when the above-mentioned comparison criteria are considered. If particular criteria are taken into consideration, the Ishikawa diagram method could be appreciated as more convenient (for example, for the simplicity, of the possibility of association the results with other research methods or of the suggestibility).

7. Conclusions

One of the analysis methods used to suggestively illustrate the causes able to generate a certain effect is the Ishikawa diagram method. In the research whose results were presented in this paper, the Ishikawa diagram method was used to highlight the factors able to affect the results of a chemical engraving process, end milling process, the 3D printing process and the factors able to determine the failure or malfunction of the computer system. Taking into consideration the industrial processes, the analysis could be made both by using the Ishikawa diagram method and the systemic analysis method. A comparison of the two analysis methods based on the use of some distinct evaluation criteria was developed using the pairwise comparison matrix and the determination of the coefficients of importance. In this way, as a global image, the systemic analysis could be preferred, but if particular criteria are applied, the Ishikawa diagram method could be appreciated as more convenient. In the future, the idea is to develop some detailed research for each process and to practically analyze how the issues during the manufacturing process can be solved and also to improve the process on the base of the theoretical premises.

8. References

[1] Ilie G, Ciocoiu CN 2010 *Management Research and Practice* **2** 1 1-20
[2] Romo T, Vick N, Quilizapa L 2013 *Fishbone Diagram & The 5 Whys* Available at http://publichealth.lacounty.gov/qiap/docs/Topic3-Fishbone.pdf Accessed: 13.09.2019
[3] Magdum RS, Gore PN 2015 *IJSRD* **3** 765-768
[4] Luca L 2016 *20th Innovative Manufacturing Engineering and Energy Conference (IManEE 2016)* A reference. IOP Conf. Series: Materials Science and Engineering **161** 012099
[5] Luo T, Wu C, Duan L 2018 *J. Clean. Prod.* **174** 296-304
[6] Slătineanu L, Nagîţ G, Dodun O, Coteaţă M, Chinesta F, Gonçalves-Coelho A, Pamies Teixeira J, San Juan M, Santo L, Santos F 2004 *Non-traditional manufacturing processes* (Chişinău: Publishing House Tehnica Info)
[7] Băloiu LM, Frăsineanu I 2001 *Management of innovation* (in Romanian) (Bucharest: Editura Economică)
[8] *Creativity methods and techniques*, 2017. Available at http://www.stiucum.com/management/managementul-cercetarii-si-dezvoltarii/Metode-si-tehnici-de-creativit81688.php Accessed: 10.08.2019
[9] Slătineanu L 2019 *Fundamentals of scientific research* (in Romanian) (Iaşi: Publishing House PIM)
[10] Belous V 1992 *Inventics* (in Romanian) (Iaşi: Publishing House Asachi)

The influence of additive manufacturing parameters on the structural and mechanical properties of acrylonitrile butadiene styrene (ABS) parts produced by fused filament fabrication

D Cristea, M A Pop, C Faraian and D Munteanu

Materials Science Department, Transilvania University of Brasov, Romania

daniel.cristea@unitbv.ro

Abstract. The key to the operation of additive manufacturing is that the parts are made by adding the material layer by layer. Each layer is a thin cross section of the part derived from the original CAD data. Each layer must have a finite thickness and thus the resulting part will be an approximation of the original data. The thinner each layer, the more the resulting piece will be more like the designed part. All additive manufacturing techniques marketed so far use a layered approach. However, certain drawbacks were reported for this type of manufacturing, all depending on certain process parameters. One of the most important build parameters, in terms of cost and final properties, is the build construction angle. The influence of the build positioning (i.e. angle of the layers) in respect to the building plate, during fused filament fabrication of acrylonitrile butadiene styrene (ABS) parts was studied. The characteristics of interest were: the structural features of the manufactured part and the mechanical properties obtained after tensile, compression, and hardness tests, all as function of the build orientation, while keeping the remaining manufacturing parameters identical. It was observed that the build orientation has a significant influence on the properties.

1. Introduction

Additive manufacturing (AM) technologies have developed rapidly and are now extensively used for the production of final parts in various industries such as aerospace, medicine, electronics, etc. [1]. Various processes were developed in order to obtain parts through the "layer-by-layer" method, from the stereolithography technique (SLA), to selective laser sintering (SLS), and beyond. One of the most widely used techniques, adopted in the present research as well, is filament fused fabrication (FFF) [2]. Acrylonitrile butadiene styrene (ABS) is used on most FFF devices but the range of available materials is very wide, including polyamides, polycarbonate, polystyrene, polylactic acid (PLA), etc. [3]. The research in the field of additive manufacturing is nowadays focused on the optimization of the processes and the involved parameters [4-5]. However, while most articles are related to understanding the process itself, to date, the literature is relatively limited regarding the influence of the print orientation on the properties of the manufactured part. When the build orientation is brought to discussion, some authors focus on the influence upon production time, waste reduction, etc. and not on how the parts' properties are affected by the printing orientation [6]. However, some reports can be found in the literature, which highlight the influence of the process parameters, including the build orientation, on the mechanical properties of polymer-based 3D printed samples. Casavola et al. report the effect of the raster parameters on the mechanical properties of ABS and PLA 3D printed parts. It was observed that the specimens showed a reduction of the mechanical characteristics if the raster angle was increased [7]. Ahn et al. [8] determined the effects of air gap, bead width, raster orientation and ABS colour on tensile and compressive strengths. The results showed that air gap and raster orientation have an important effect on the tensile strength. On the other hand, the other parameters have negligible effects. Byberg et al. report the study of the impact of the two process parameters, layer orientation and build direction, on the mechanical properties of the ULTEM 9085 material processed by fused deposition modelling. The authors found that the printing direction that gives the best results is the edge direction. This build direction showed a high level of strength for all the tests performed. In this particular case, the time for printing in this direction is somewhat greater than for flat samples, but less material would be used [9]. Zaldivar et al. investigated the print orientation effects on the macrostructure, the mechanical and thermal properties, and the strain field behaviour of the same ULTEM 9085 material, using Fused Deposition Modelling as the printing technique. In this particular case it was observed that increasing the angle from the build platform results in microstructures that further reduce the volume fraction of extruded fibre material from the primary load direction, resulting in lower strength [10]. The fracture resistance and interlayer adhesion of ABS 3D printed parts was studied by Nahal Aliheidari et al., observing that the samples printed at higher temperatures exhibited fracture resistance close to the fracture resistance of bulk ABS, indicating excellent interlayer bonding, as function of the extrusion temperature [11]. A comprehensive review regarding the influence of build parameters on the mechanical behaviour of polymeric 3D printed parts suggests that the main parameters responsible for adequate characteristics are: raster-to-raster air gap, raster angle, layer thickness, infill density and build orientation [12]. However, the same authors suggest that the intention to generalize results from the literature would be impossible, mostly due to the issue of intra-3D printer variability. This means that one could not implement ideal build parameters obtained on a certain machine/polymer/application combination, onto a different printer. Consequently, our aim was to add to the already available knowledge on the subject of mechanical properties of 3D-printed ABS parts, by studying the influence of the build orientation on the tensile and compressive stress resistance of ABS samples produced on an FFF/FDM machine (CreatBOT DX 3D printer).

2. Sample preparation and analysis techniques

2.1. Sample preparation

The samples were manufactured on a CreatBOT DX 3D printer, at 290 °C printing temperature and a 200 μm layer thickness. The filament used in the construction of the samples is made of ABS (Sakata 3D supplier), with a diameter of 3 mm, and it was extruded through a nozzle with a diameter of 0.6 mm. Even though the recommended extrusion temperature given by the manufacturer is between 235-256 °C, when the actual printing of the specimens was initiated, several defects occurred during this process (delamination of the layers). Consequently, the temperature was gradually increased to 290 °C, the temperature at which the specimens were obtained at the desired quality. In order to assess the influence of the build construction angle on the properties, the samples were positioned as described in figure 1. One can observe the tensile test samples, positioned on the edge and flat sides, in respect to the building plate, as well as the compression samples (cylindrical blocks), positioned horizontally, vertically and at a 15-degree angle in reference to the building plate. The dimensions of the samples were: i) tensile test samples: total length 94 mm, calibrated length 50 mm, thickness 3 mm, radius between the calibrated section and the fastening sections 4 mm, calibrated width 4 mm, fastening section width 16 mm; ii) cylindrical blocks for compression tests: diameter 10 mm, height 15 mm. Each configuration of the samples was built in five pieces, for statistical relevance of the results after testing.

Figure 1. The positioning of the ABS samples during printing.

2.2. Sample analysis

The tensile and compression tests were performed on a WDW-150S universal testing machine, according to ASTM D638 – 14 (tensile), and ASTM D695 – 15 (compression) standards. Sample pieces cut from the tensile test parts and the compression blocks were embedded in resin and polished, in order to analyse the placement and characteristics of the layers, through optical microscopy. The polishing process was performed on a Labo-Pol 5 polishing machine, from Struers, using five granulations, from 800 to 5000 grit. The samples were further analysed on an optical microscope (Nikon Eclipse MA100), with various magnifications. Furthermore, the hardness of the specimens was assessed using a Shore D hardness tester (Sauter). For statistical relevance, the tensile and compression tests were performed five times, while the Shore D hardness measurements were performed ten times on each sample, and the values were averaged. In order to assess the thermal stability of the ABS filament, Differential Scanning Calorimetry (DSC) analysis has been performed in the -100 ÷ +450 °C domain under nitrogen atmosphere, with 10 °C/min cooling/heating rate (DSC-200 F3 Maia/STA 449F3 Jupiter, Netzsch).

3. Results and discussions

The variation in tensile breaking strength and of the Shore D hardness (after the tensile tests), as function of the build orientation can be observed in figure 2. The variation of stress as function of strain can be observed in figure 3. Both the tensile breaking strength and the Shore D hardness exhibit larger values for the flat printed samples, namely 32.6 MPa and 77 HD, while the edge printed samples exhibit slightly poorer characteristics (29.2 MPa and 74 HD). This phenomenon could be related to the internal structure of the samples, which can be observed in figure 4. The flat printed sample is characterized of structural homogeneity, with a significantly lower number of defects, compared to the edge printed sample. The edge printed sample exhibits a great number of pores, caused by the loss of adhesion/fusion between the rows of ABS. The influence of air-gap on the tensile strength was shown to have a detrimental effect [8]. This phenomenon could be related to the build orientation and to the position of each row in relation to the heating sources during printing, namely the printing head and the heated printing plate. Even if the printing head keeps the same temperature for the extrusion, the distance between the heated plate, on which the samples are built, and the upper rows of the sample is increasingly larger for the edge-printed sample, compared to the flat-printed sample. Consequently, the already built rows are cooling much faster, causing a loss of adhesion between the printed material and the incoming material.

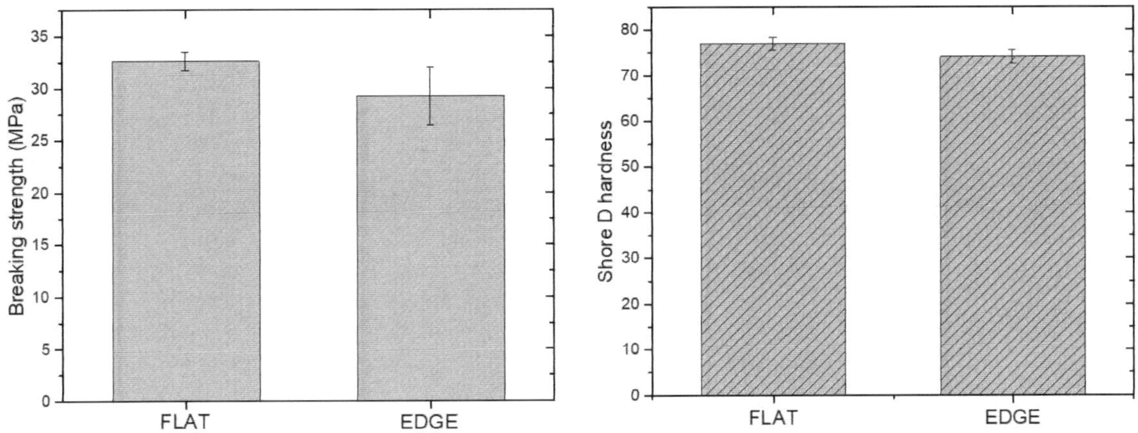

Figure 2. The variation of the tensile breaking strength and of the Shore D hardness (after testing), as function of the build orientation.

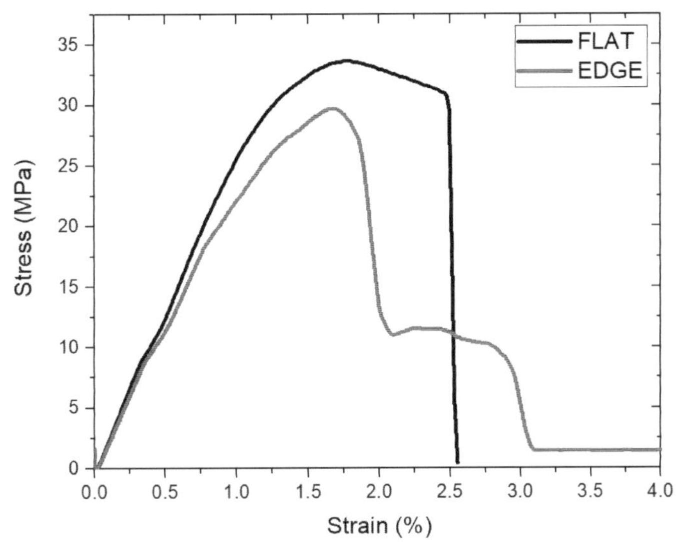

Figure 3. The variation of stress as function of strain, for the flat and edge printed samples.

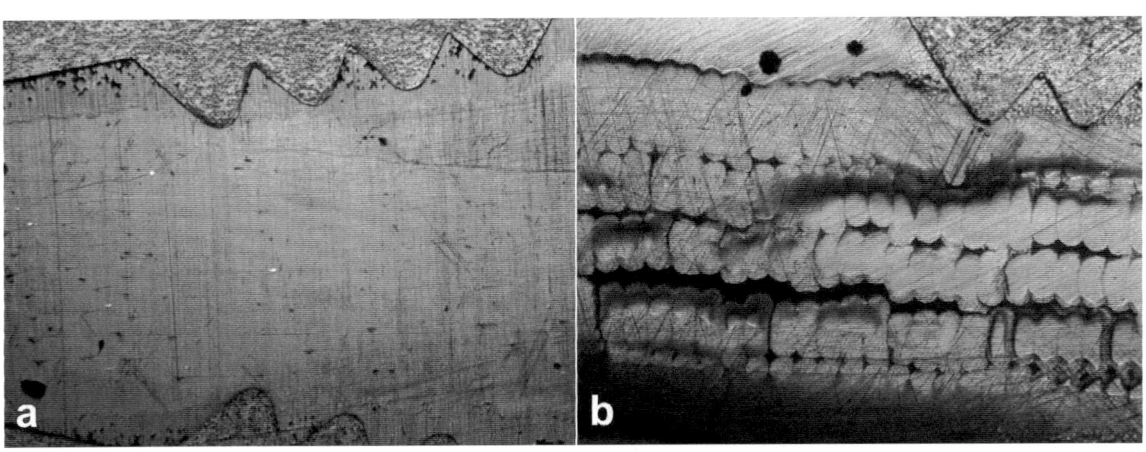

Figure 4. The internal structure and defects inside the samples, after the tensile tests:
a) the sample printed on the flat side; b) the sample printed on the edge side. (Scale length 500 μm)

As far as the compression tests are concerned, a similar variation between samples can be observed, albeit to a lesser degree, especially when discussing about the variation of the Shore D hardness, measured after the compression tests. Minimal differences should be expected in relation to the Shore D hardness values, mostly due to the fact that the material is compressed during the test, and the influence of the voids is minimized, thus leading to an increase in homogeneity of the samples. The variation of the compression strength and that of the Shore D hardness can be observed in figure 5, as function of the build orientation. Other reports suggest that the differences in compressive strength are minimal, regardless of the build orientation or other build parameters [8], thus confirming our findings. The difference between the compression strength for the vertically and horizontally built samples could be explained if one looks at the internal structure of the samples, before the compression tests, shown in figure 6. The dimension of the pores/voids is much smaller inside the vertically built samples, compared to the horizontally built samples. The internal structure of the samples built at a 15-degree angle in relation to the build plate was similar to the samples built vertically, thus it is not shown in this manuscript. Consequently, the compression strength for these two samples is relatively similar (52.75 MPa for the vertically built, versus 51.5 MPa for the 15-degree built sample), especially considering the overlapping measuring error.

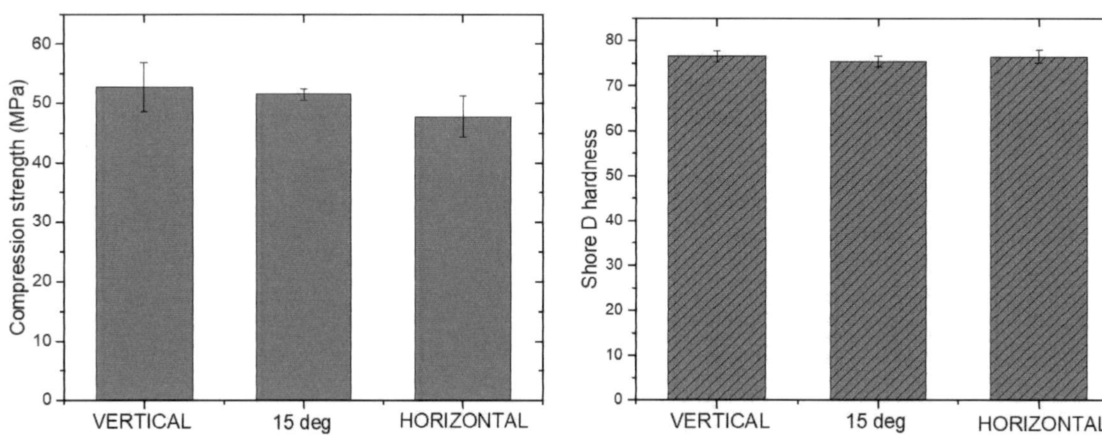

Figure 5. The variation of the compression strength and of the Shore D hardness (after testing), as function of the build orientation.

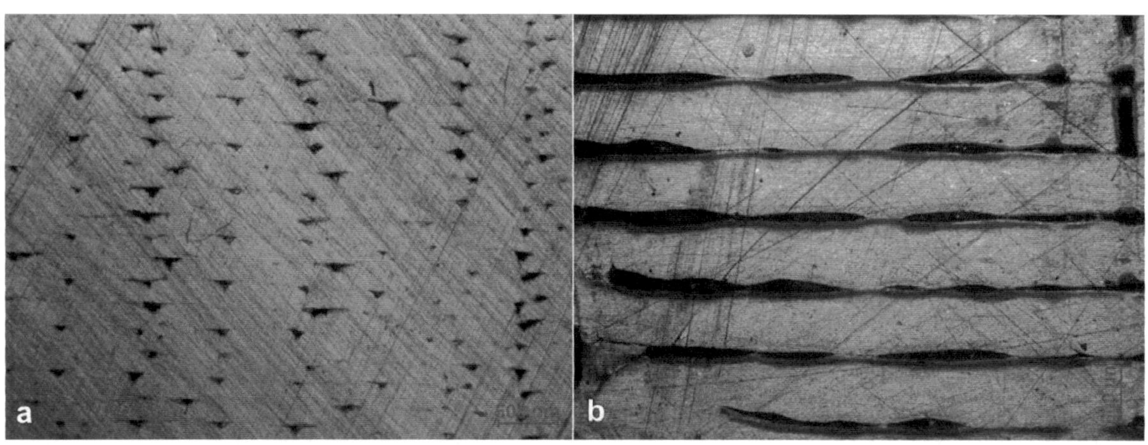

Figure 6. The internal structure of the samples, before compression testing, on the as-built samples: a) vertical sample; b) horizontal sample.

The internal structure of the samples after the compression tests can be observed in figure 7. One can observe that the internal voids disappeared almost entirely, to a higher degree inside the vertically and 15-degree built samples, and to a lesser degree inside the horizontally built sample. Moreover, one can observe the shifting from the vertical symmetry axis of the lateral void locations, due to the barrelling effect caused by the friction force which acts at the contact surface between the compression plates and the sample, during the compression test. Moreover, the thickness of the individual layers, as expected, is smaller as compared to the thickness prior to the compression tests. The presence of the voids inside the as-printed samples seems to have an influence even on the mass of the final part, as function of the build orientation. The estimated values of the print duration and mass of the final part, calculated for a printing speed of 50 mm/sec are as follows: for the flat-built tensile test samples, 116 minutes with a mass of 31 g, while the edge-built tensile test samples, 113 minutes with a mass of 29 g; the horizontally-built compression samples, 26 minutes, with a mass of 7 g, the 15-degree built samples, 29 minutes with a mass of 7 g, while the vertically-built sample, 30 minutes, with a mass of 7 g. These values are of practical importance when choosing either the mechanical properties needed for a certain application, or the economy in terms of building time or final mass.

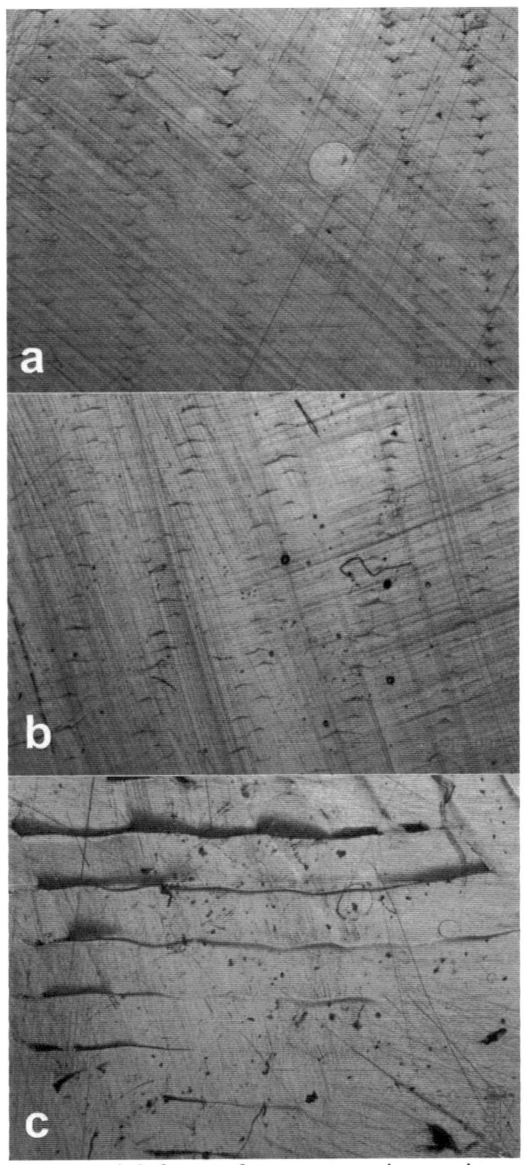

Figure 7. The internal structure and defects, after compression testing, as function of the build orientation: a) vertical, b) 15 degrees inclination in relation to the build plate, c) horizontal.

The thermogram from figure 8 represents the thermal stability of the ABS filament, when subjected to heating. One can observe a "softening" temperature, related to the endothermic styrene glass transition at ~140 °C, and endothermic acrylonitrile glass transition temperature, between 200-220 °C, while melting occurs at 420-424°C. Since the rubber (butadiene) phase of ABS is susceptible to thermal degradation, crosslink points are formed between the macromolecular chains, which lead to a decreased chain mobility and an increase in the melting temperature is expected for the heat processed (extruded) print. The temperature of interest from this thermogram is the one related to the acrylonitrile glass transition, between 200-220 °C, which is relatively close to the recommended printing temperature, as per the filament manufacturer data sheet. It would seem that the acrylonitrile glass transition occurs in a relatively larger domain than expected, thus potentially leading to the printing defects observed during the initial print runs, and to the need for the filament extrusion temperature increase, from 235-256 °C to 290 °C.

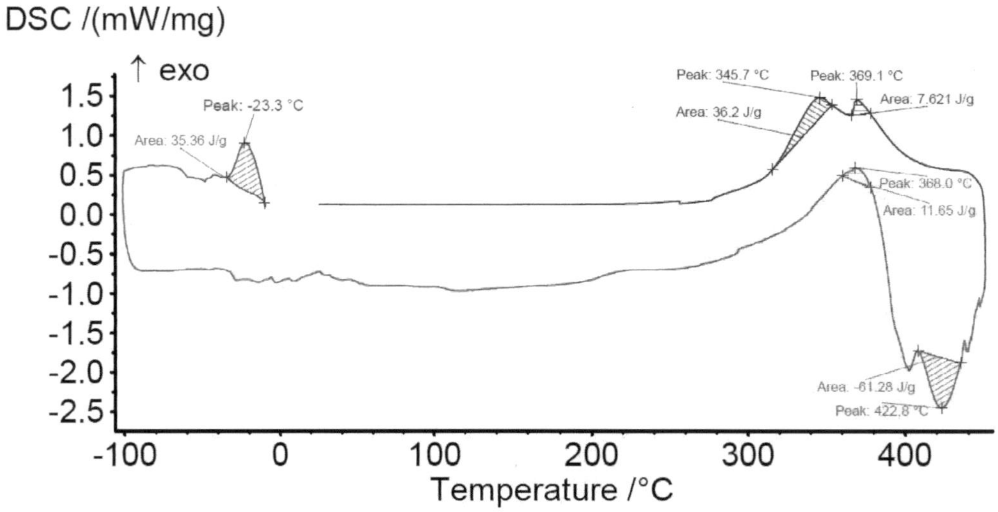

Figure 8. The thermal stability of the ABS filament, obtained from differential scanning calorimetry analysis.

4. Conclusions

Acrylonitrile butadiene styrene (ABS) parts were produced by fused filament fabrication, with identical printing parameters, while changing the build orientation in relation to the building plate. Firstly, a significantly larger extrusion temperature was needed, compared to the filament manufacturer data sheet, to be able to obtain structurally sound samples (290 °C, compared to 235-256 °C). Secondly, the mechanical properties were clearly influenced by the build orientation, the tensile test samples built on the flat side showed increased strength compared to the edge-built samples, while the vertically-built compression test samples showed increased strength compared to the 15-degree built and horizontally built samples. Marginal differences were noticed in terms of Shore D hardness variation, mostly related to the internal structure after testing. The appearance of voids is linked to the temperature at the building site, larger distances from the building plate, for identical extrusion temperature, leading to a greater number of voids.

5. References

[1] Stansbury J and Idacavage M 2016 3D printing with polymers: Challenges among expanding options and opportunities *Dent. Mater.* **32** 54–64
[2] Balletti C, Ballarin M and Guerra F 2017 3D printing: State of the art and future perspectives *J. Cult. Herit.* **26** 172–182
[3] Guo N and Leu M C 2013 Additive manufacturing: technology, applications and research needs *Front. Mech. Eng.* **8** 215–243

[4] Rodriguez J, Thomas J and Renaud J 2003 Mechanical behavior of acrylonitrile butadiene styrene fused deposition materials modeling *Rapid Prototyp.* **9** 219–230
[5] Hebda M, McIlroy C, Whiteside B, Caton-Rose F and Coates P 2019 A method for predicting geometric characteristics of polymer deposition during fused-filament-fabrication *Addit. Manuf.* **27** 99–108
[6] Zwier M and Wits W 2016 Design for additive manufacturing: Automated build orientation selection and optimization *Procedia CIRP* **55** 128 – 133
[7] Casavola C, Cazzato A, Moramarco V and Pappalettere C 2016 Orthotropic mechanical properties of fused deposition modelling parts described by classical laminate theory *Mater. Des.* **90** 453–458
[8] Ahn S-H, And S R, Wright P K, Montero M, Odell D and Roundy S 2002 Anisotropic material properties of fused deposition modelling ABS *Rapid Prototyp. J.* **8(4)** 248–257
[9] Byberg K I, Gebisa A W and Lemu H G 2018 Mechanical properties of ULTEM 9085 material processed by fused deposition modelling *Polym. Test.* **72** 335–347
[10] Zaldivar R J, Witkin D B, McLouth T, Patel D N, Schmitt K and Nokes J P 2017 Influence of processing and orientation print effects on the mechanical and thermal behavior of 3D-Printed ULTEM 9085 Material *Addit. Manuf.* **13** 71–80
[11] Aliheidari N, Tripuraneni R, Ameli A and Nadimpalli S 2017 Fracture resistance measurement of fused deposition modeling 3D printed polymers *Polym. Test.* **60** 94–101
[12] Popescu D, Zapciu A, Amza C, Baciu F and Marinescu R 2018 FDM process parameters influence over the mechanical properties of polymer specimens: A review *Polym. Test.* **69** 157–166

Acknowledgments
This work was supported by a grant of the Romanian National Authority for Scientific Research and Innovation, CNCS/CCCDI - UEFISCDI, project number PN-III-P3-3.6-H2020-2016-0077, within PNCDI III. We hereby acknowledge the structural founds project PRO-DD (POS-CCE, O.2.2.1., ID 123, SMIS 2637, ctr. No 11/2009) for providing some of the infrastructure used in this work.

Aluminum based metastable alloys for additive manufacturing

A Nitoi, D Cristea, M A Pop, T Bedo, B Varga and D Munteanu

Department of Materials Science, Transilvania University of Brasov, 500036 Brasov, Romania

alexandra.nitoi@unitbv.ro

Abstract. Additive manufacturing (AM) is a technology which has the potential to replace conventional casting methods. Its capacity to obtain complex shapes is the main advantage over other manufacturing techniques. Metastable structures in the form of supersaturated or amorphous/nano solid solutions could potentially help the manufacturing process, due to energy release during heating, caused by the phase transformations towards stable structures. Aluminum-based alloys are of interest due to their combined high specific strength to weight ratio, good ductility and superior corrosion resistance. The paper presents results concerning the development of Al-Si-Ni alloys in bulk and melt-spun ribbon forms. The thermal stability of the aluminum-based alloys was analyzed by Differential Scanning Calorimetry, while some of the mechanical properties were assessed by instrumented indentation. The fast-cooled alloys exhibit metastable structures, inferred from the exothermal peaks observed during heating. A hardening phenomenon was also observed in case of the melt-spun ribbons, compared to the bulk alloys.

1. Introduction

In contrast to conventional processing methods, additive manufacturing (AM) is based on an incremental layer-by-layer manufacturing. Nowadays, parts with better properties and intricate geometries can be additively manufactured more easily compared to the traditional subtractive methods. To date, only a handful of metallic materials are used in AM. Consequently, the development of structural materials with enhanced mechanical properties is of great interest in the additive manufacturing field [1]. Aluminum alloys are used in the aerospace and automotive industries because they exhibit attractive properties: excellent oxidation resistance, high specific strength, high thermal and electrical conductivity, low density and corrosion resistance [2]. One of the most important aluminum-based alloys is the Al–Si group of alloys, due to their light weight and good mechanical properties [3]. Adding nickel to the Al-Si alloys could potentially lead to improved properties. Moreover, certain compositions of the Al-Si-Ni class of alloys are relatively easily obtained in amorphous or nanocrystalline structures, through fast cooling techniques. The description of the phase equilibrium of the ternary Al–Ni–Si system is far from being complete, although considerable work has been done already [4, 5, 6]. Karthik Chandrasekaran et al. analyzed certain compositions from this class of alloys, in accordance with the ternary phase equilibrium diagram, with X-ray diffraction (XRD), metallography, electron probe microanalysis (EPMA), and differential thermal analysis (DTA) techniques, and determined the various invariant reactions within the three sections at 70, 75 and 80 at. % Ni. Karthik observed large solubility of Al and Si in Ni [5]. Ni-based alloys have high melting point, good oxidation resistance and excellent high-temperature mechanical properties [7]. Some of the ductile Al-based amorphous alloys have been found to exhibit superconductivity [8]. Amorphous materials or nanocrystalline materials could exhibit some phase transformations upon heating and cooling, measurable as exothermal or endothermal peaks. It can be hypothesized that certain exothermal transformations could contribute to a more efficient additive manufacturing process, due to a lower melting point of the metallic powders. The present paper details experimental results concerning Al-Si-Ni metastable alloys, obtained in melt-spun ribbon form, compared to conventional castings of Al-Si-Ni. The results of interest were their thermal stability and hardness.

2. Materials and methods

The chosen compositions of alloys were $Al_{66}Si_{18}Ni_{16}$, $Al_{68}Si_{18}Ni_{14}$, $Al_{70}Si_{18}Ni_{12}$. These compositions were obtained by casting, after their elaboration in a silicon carbide resistor furnace, under a protective flux layer to avoid oxidation. The molten alloys were poured in steel crucibles. These preliminary alloys are referred hereinafter as "bulk". Thin alloy ribbons from these same compositions were obtained by melt-spinning. This process facilitates obtaining very high cooling rates and the possibility to obtain amorphous or nanocrystalline materials. The schematic of the melt spinning technique is presented in figure 1. The molten alloy, heated by an induction coil, is ejected from the crucible, due to an applied pressure, through a small orifice (1 mm diameter), and it comes in contact with a rotating copper wheel, which serves as the cooling/solidification medium. In order to assess the influence of the cooling rate on the thermal stability of the melt-spun ribbons, two rotation speeds for the copper cooling wheel were used, namely 32 m/s and 15 m/s. Consequently, nine sample configurations were obtained: as-cast bulk samples, melt-spun ribbons obtained with a lower cooling rate, and melt-spun ribbons obtained with a higher cooling rate. The chemical compositions were chosen according to the ternary Al-Si-Ni equilibrium phase diagram (shown in figure 2), from the Al-rich corner. Higher nickel and silicon contents usually leads to brittle compositions, hence the reason behind choosing Al-rich compositions.

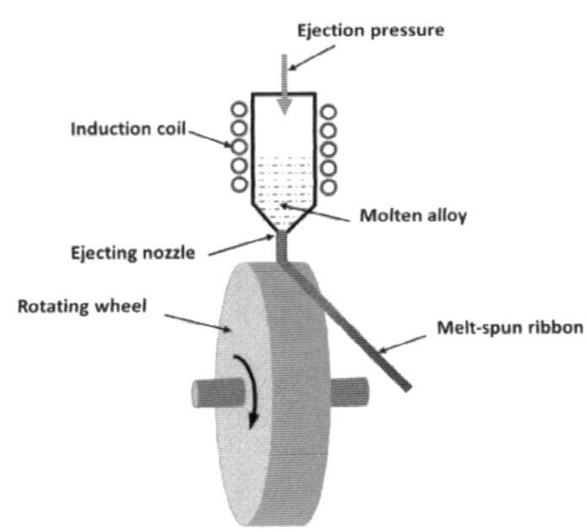

Figure 1. Melt spinning process diagram [9]

The thermal stability was assessed by thermal analysis, namely Differential Scanning Calorimetry (DSC), using a DSC 200 F3 Maia apparatus. Tests were carried out in the maximum temperature range, from room temperature (RT) to 600 °C, in nitrogen atmosphere (to avoid oxidation), with a heating/cooling rate of 10 °C/min. The data processing was done using the NETZSCH Proteus analysis software.

The hardness of the melt-spun ribbons was assessed by instrumented indentation. This technique differs from classical macrohardness measurements, where the impressions are first generated by the indenter, and then imaged using a microscopy technique. In the case of instrumented indentation, the load and penetration depth are simultaneously recorded during both the loading and unloading stages, resulting in a load–displacement diagram. A maximum applied load of 20mN, loading rate 60 mN/min, unloading rate 180 mN/min, and dwell period of 2s (to minimize the creep effect) were used.

10 indentations were performed on each sample, and the results were averaged. The melt-spun ribbons were embedded in low-contraction resin, followed by mechanical polishing on a Struers Labo-Pol machine, to a mirror-like finish.

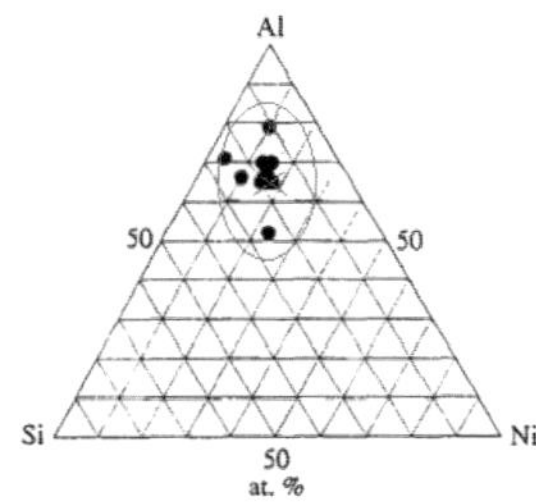

Figure 2. Chemical compositions position from the Al-Si-Ni system, which can be relatively easily obtained in amorphous or nanocrystalline form.

3. Thermal stability and instrumented indentation results

The thermal stability of the processed alloys is shown in figures 3 and 4. Figure 3 represents the variation of DSC as function of the temperature, where samples with the same chemical composition, but with different cooling rates are compared. One can observe that none of the bulk samples exhibit any exothermal peaks. The only events which are occurring are endothermal phenomena, which can be attributed to the melting point of the eutectic phase from the alloys, at approximately 568 °C.

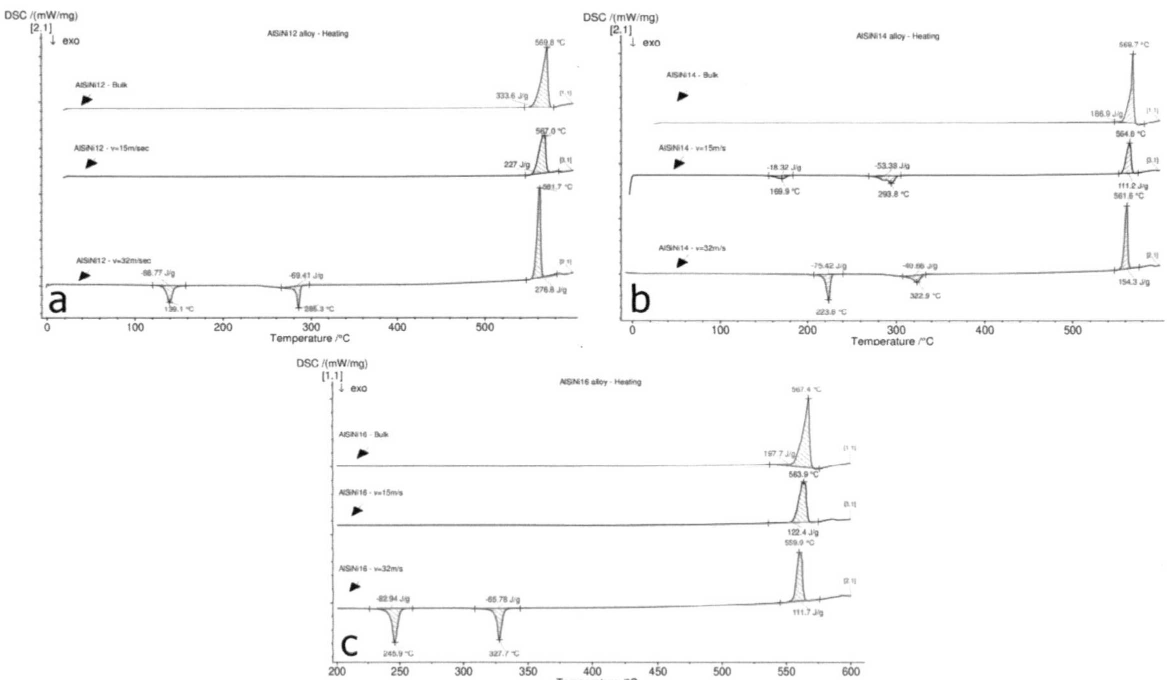

Figure 3. Differential scanning calorimetry curves, as function of the temperature: (a) $Al_{70}Si_{18}Ni_{12}$; (b) $Al_{68}Si_{18}Ni_{14}$; (c) $Al_{66}Si_{18}Ni_{16}$ - bulk, v=15m/s, v=32m/s

Contrastingly, the melt spun ribbons exhibit exothermal peaks upon heating, thus showing their metastable nature. However, certain differences between samples can be observed. The most striking difference is represented by the fact that only the $Al_{68}Si_{18}Ni_{14}$ composition exhibits exothermal peaks for the lower cooling rate, inferred from the lower rotation speed of the copper cooling wheel. The influence of the rotation speed on the thermal stability is shown in Figure 4. One can observe that the peak intensity is significantly enlarged for the melt-spun ribbons obtained higher rotation speed.

According to the Al-Si-Ni equilibrium phase diagram, three potential phases can be formed in these alloys: the eutectic, and the Al_3Ni and $Ni_2(Al_{1-x}Si_x)_3$ intermetallic compounds. It would seem that with higher nickel content, these exothermal peaks are located at higher temperatures. These exothermic transformations could reduce the energy consumption required for the melting and sintering of alloy granules during additive manufacturing processes.

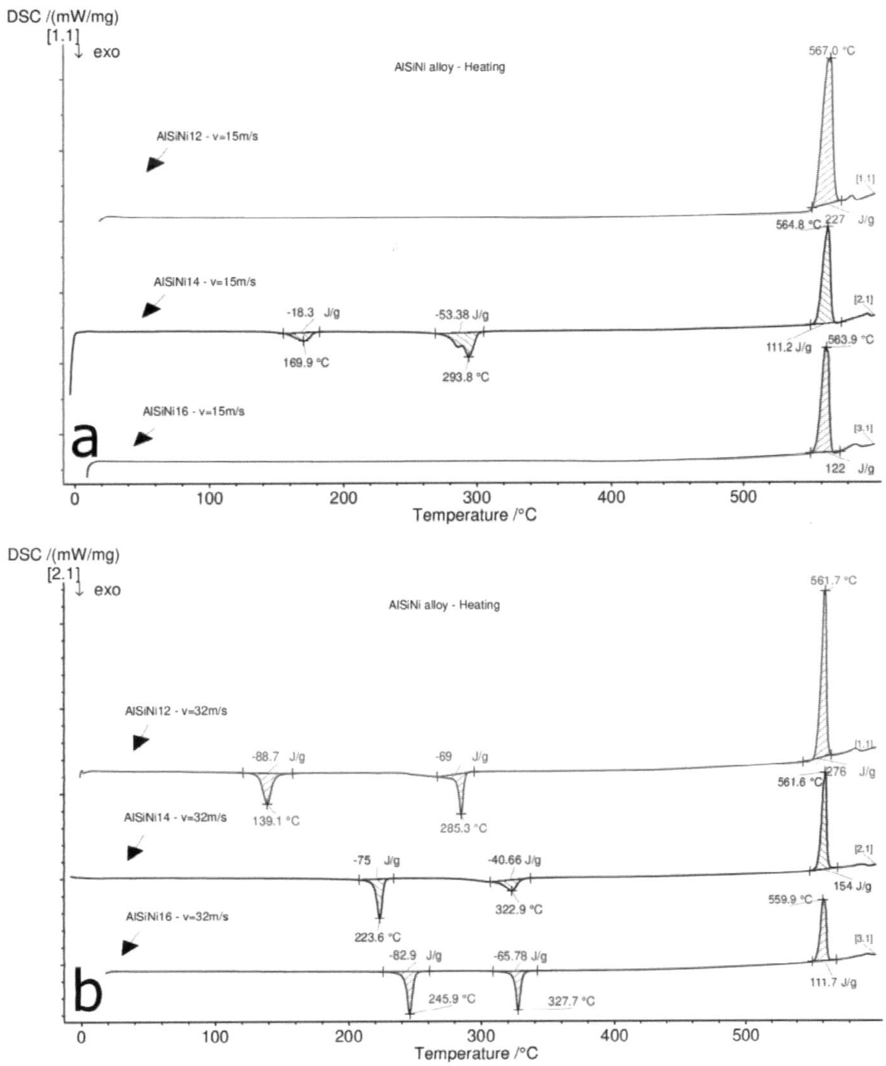

Figure 4. Differential scanning calorimetry analysis comparison, as function of the chemical compositions, for the melt-spun ribbons: (a) $Al_{70}Si_{18}Ni_{12}$; $Al_{68}Si_{18}Ni_{14}$; $Al_{66}Si_{18}Ni_{16}$ - v=15 m/s; (b) $Al_{70}Si_{18}Ni_{12}$; $Al_{68}Si_{18}Ni_{14}$; $Al_{66}Si_{18}Ni_{16}$ - v=32 m/s

In figure 5 the loading-unloading curves obtained on the melt-spun ribbons, processed with two cooling rates (15 m/s and 32 m/s) are presented. The hardness of a material is directly related to the maximum penetration depth obtained for a certain load. From this observation one can infer that, with the exception of the $Al_{68}Si_{18}Ni_{14}$ melt-spun ribbon, obtained with a peripheral speed of v = 15 m/s, higher cooling rates are leading to increased hardness. Furthermore, the effect of the cooling rate on the structural development is shown in figure 6. The samples obtained with 32 m/s rotation speed (Figure 6 a, c, e) are featureless, with no visible crystalline regions. A certain degree of oxidation due to the polishing procedure is evident, especially on sample $Al_{68}Si_{18}Ni_{14}$ v = 32 m/s. Using a lower rotation speed (v = 15m/s, Figure 6 b, d, f), thus a lower cooling rate, leads to the formation of crystalline grains, separated into domains which exhibit different mechanical responses when the diamond indenter was pressed on the material, a smaller area of the indentation imprint signifying a higher hardness.

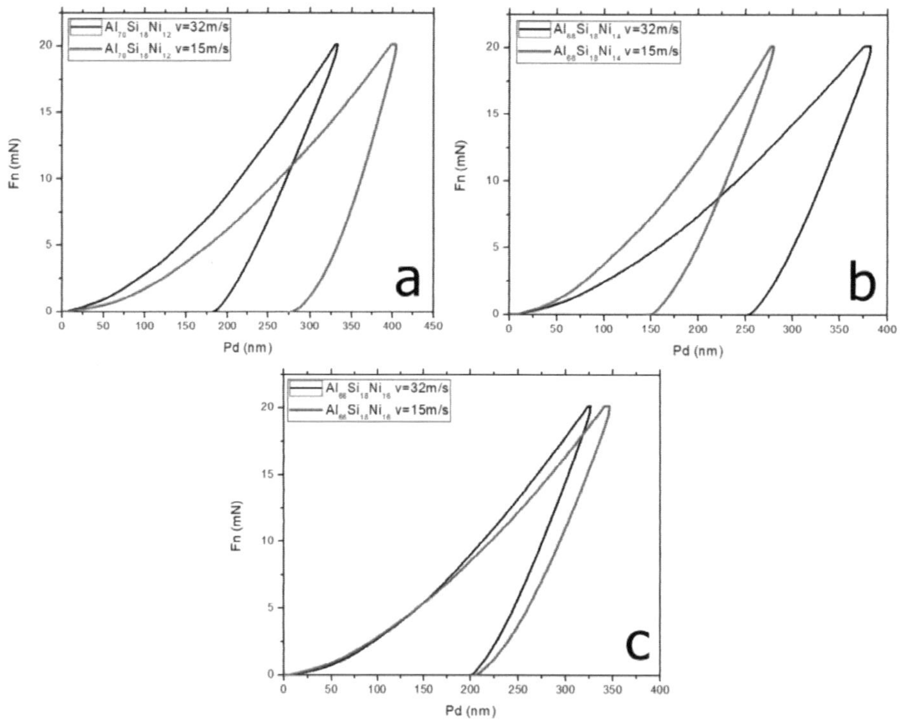

Figure 5. Instrumented indentation loading-unloading curves, compared as function of the rotation speed: (a) $Al_{70}Si_{18}Ni_{12}$; (b) $Al_{68}Si_{18}Ni_{14}$; (c) $Al_{66}Si_{18}Ni_{16}$ (Fn – applied load; Pd – penetration depth)

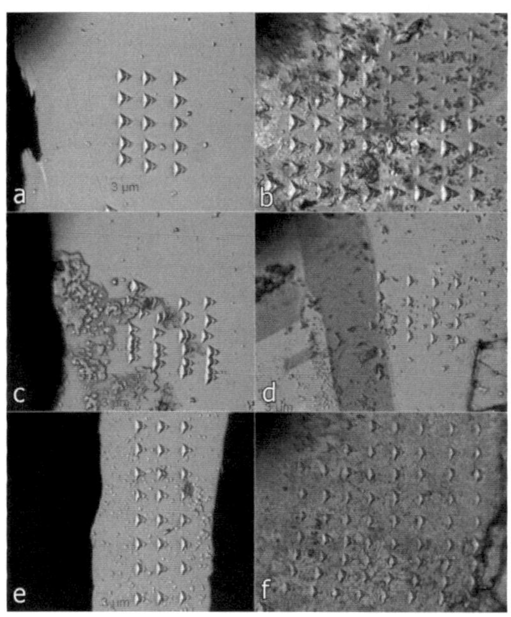

Figure 6. Indentation imprints on the melt-spun ribbons: a) $Al_{70}Si_{18}Ni_{12}$ v = 32 m/s; b) $Al_{70}Si_{18}Ni_{12}$ v = 15 m/s; c) $Al_{68}Si_{18}Ni_{14}$ v = 32 m/s; d) $Al_{68}Si_{18}Ni_{14}$ v = 15 m/s; e) $Al_{66}Si_{18}Ni_{16}$ v = 32 m/s; f) $Al_{66}Si_{18}Ni_{16}$ v = 15m/s

4. Conclusions

In this work, the thermal stability and indentation hardness of certain compositions from the Al-Si-Ni alloy system were investigated. The fast-cooled samples were obtained by melt spinning, and the analysis results were compared to the ones on samples obtained by gravity casting in steel crucibles. From the experimental results, rapid solidification resulted in a significant structural refinement, compared to the alloys processed by classical methods. Four out of six sample configurations exhibited exothermal phase transformations, which could benefit additive manufacturing processes. The exothermal peak intensities were increased for the samples obtained with a higher cooling rate, inferred from a higher rotation speed of the copper cooling wheel. Moreover, increased resistance to plastic deformation was observed for the samples obtained with higher cooling rates. The structural refinement was observed on the optical micrographs especially for the samples obtained with 32 m/s rotation speed. The remaining samples exhibit relatively large crystalline domains, each domain showing different mechanical response, inferred from the difference in size of the indentation imprints. Considering AM technologies, the ribbons obtained by melt-spinning with 32 m/s rotation speed could potentially reduce the energy consumption during manufacture.

5. References

[1] Li N, Huang S, Zhang G, Qin R, Liu W, Xiong H, Shi G and Blackburn J 2019 Progress in additive manufacturing on new materials: A review *J. Mater. Sci. Technol.* **35** 242–269
[2] Croteau J R, Griffiths S, Rossell M D, Leinenbach C, Kenel C, Jansen V, Seidman D N, Dunand D C and Vo N Q 2018 Microstructure and mechanical properties of Al-Mg-Zr alloys processed by selective laser melting *Acta Mater.* **153** 35–44
[3] Chen C-L, Richter A and Thomson R C 2009 Mechanical properties of intermetallic phases in multi-component Al–Si alloys using nanoindentation *Intermetallics* **17** 634–641
[4] Richter K W and Ipser H 2003 The Al–Ni–Si phase diagram between 0 and 33.3 at.% Ni *Intermetallics* 11(2) 101–109
[5] Chandrasekaran K, Richter K W and Ipser H 2006 The Al–Ni–Si phase diagram—Part III: Phase equilibria in the nickel rich part *Intermetallics* **14** 491–497
[6] Richter K W, Chandrasekaran K and Ipser H 2004 The Al–Ni–Si phase diagram. Part II:phase equilibria between 33.3 and 66.7 at.% Ni *Intermetallics* **12(5)** 545–554
[7] Cadırl E, Herlach D M and Volkmann T 2010 Characterization of rapidly solidified Ni–Si and Co–Al eutectic alloys in drop tube *J. Non-Cryst. Solids* **356** 461–466
[8] Inoue A, Ohtera K and Masumoto T 1988 New Amorphous Al-Y, Al-La and Al-Ce Alloys Prepared by Melt Spinning *Jpn. J. Appl. Phys* **27(5)** L736–L739
[9] Rong C and Shen B 2018 Nanocrystalline and nanocomposite permanent magnets by melt spinning technique *Chin. Phys. B.* **27(11)** 117502

Acknowledgement

This project has received funding from the European Union's Horizon 2020 research and innovation programme under grant agreement No. 723699. We hereby acknowledge the structural funds project RO-DD (POS-CCE, O.2.2.1., ID 123, SMIS 2637, ctr. No 11/2009) for providing the infrastructure used in this work.

IOP Conf. Series: Materials Science and Engineering **682** (2019) 012015 doi:10.1088/1757-899X/682/1/012015

Assessing the influence of significative parameters used in 3D Fused Filament Deposition

C Pisarciuc

Transilvania University of Brasov, Romania

pisarciuc.c@unitbv.ro

Abstract. The present article presents an analysis of the several parameters used at Fused Filament Deposition-FFD 3D printing process. To evaluate these parameters, a laptop / tablet support has been chosen as a subject that presents different characteristics both from the point of view of mechanical demands, to which it is subject, as well as geometry. The reason for choosing such an object to be produced by additive technology is the fact that in its geometry are encountered most of the spatial surfaces that present difficulties to be obtained by this method. In this context, the article highlights the steps that are necessary to get 3D printed objects to meet the initial design requirements. For the assessment of the degree of influence, parameters were identified and then they were evaluated to find an optimal solution. In the conclusions are presented some useful values for practical applications of this type and some considerations about two types of 3D slicers.

1. Introduction

The 3D printers that use filament are based on additive technology for depositing successive layers of material. The material to be deposited is stored on a filament spool, which ensures the continuous feeding of the extruder. The extruder is feed with the material, heats it at a certain temperature, and places it in successive layers of a certain thickness and in a certain shape following a specific path of each layer. In other words, in this process, a thermoplastic filament is extruded through an extruder and deposited into successive layers [1, 2].

The main factor differentiating printers in the additive manufacturing (AM) category is the material used to obtain the final body. This article will provide examples for a Fused Filament Deposition (FFD) printer. In some cases, in order to avoid the use of registered names, the Layer Plastic Deposition - LPD is also used to characterize the used technology. Unfortunately, this latest acronym, although it better describes technology, is also used in other areas.

2. Printer parameters and object characteristics

The first step in obtaining an object through fused filament fabrication additive technology is to configure the printer's parameters, which is done according to the printer's characteristics, the type of material used, the shape and characteristics of the final object. Configuration is done using specific software. In this case, it was used Z-Suite [3] and Slic3r [4] software. The Z-suite is printer's proprietary software and the process is much simplified. In contrast, with Slic3r the configuration is more laborious, but not complicated.

2.1. Object properties

As a subject to be 3D printed 3D by additive technology is a laptop / tablet support as shown in the figure 1.

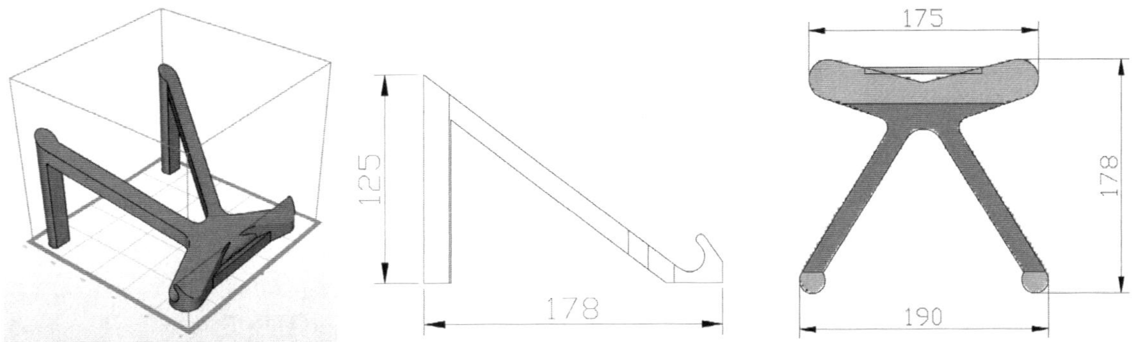

Figure 1. Laptop / tablet support and its dimensions.

The dimensions and shape of the support were chosen to accommodate either a laptop or a tablet under static equilibrium conditions. The reason for choosing such an object to be produced by additive technology is the fact that in its geometry are encountered most of the spatial shapes that present difficulties to be obtained by this method.

Considering the destination, it is necessary that the material from which the support is built can maintain its properties at relatively high temperatures under stiffness conditions. The material that meets both conditions simultaneously is the polylactic acid PLA type [5]. PLA is the most common 3D printing material because it is easy to use and is made from renewable resources and thus, biodegradable. PLA filament is useful in a broad range of 3D printing applications, being both odourless and low-warp, and does not require a heated bed [6, 7].

Although object orientation is not a configuration parameter or a print parameter, it plays a very important role in choosing the other print parameters. Specifically, depending on the spatial geometry of each printed object, it is necessary to determine the optimal orientation of the deposited layers. The support surface of the object on the depositing tray is recommended to be as large as possible. This mode of settlement simultaneously generates an increased stability and good adhesion to printing bed. The layers that make up the inclined surfaces outside the previously deposited layer face difficulties in obtaining them with respect to the planar surfaces. For the above reason, in order to avoid aesthetic and precision defects, it is necessary for the printed object to exhibit as few inclined surfaces as compared to the previously deposited state. The optimal orientation of the support is presented in first picture from figure 1.

2.2. Dimensions of the build platform

This project will use a Zortrax M200 [8] 3D printer. The dimensions of the print bed define the maximum distances traversed by the print head along the X and Y axes and are specific to each individual printer. For mentioned printer is a 200 mm square. Considering this form, the origin can be set in the lower left corner.

2.3. Diameter of the hot end

The hot end diameter determines the maximum height of the deposited layer (on Z-axis) and its width (on XY plane). The minimum height of the deposited layer is also determined by the positioning accuracy of the extruder on the Z-axis. Opposite, the maximum height cannot be larger than the diameter of the nozzle. The width of the deposited layer correlates with the diameter of the nozzle but also depends on the extrusion speed (usually 120% of nozzle diameter) [9]. In this context, the extrusion speed is the amount of material that the print head can deposit in the time unit.

2.4. Filament diameter
Specifically, there are several diameters available for the filaments used. Usually, there are no major differences in the results obtained depending on the diameter used but the diameter of the filament must be correlated with the diameter of the nozzle. For this support, a 1.75 mm filament was chosen.

2.5. Printing temperature
The extrusion temperature is the temperature reached by the hot end during deposition. This temperature is extremely important in relation to the quality of the printed object. The main influence of the extrusion temperature is exerted on the final surface roughness of the object. As a rule, PLA materials have a melting point in the range of 160 ~ 230 degrees Celsius and the value chosen to achieve the support is 200° C.

3. Process parameters
Configuration of print parameters is a necessary step because extrusion parameters depend on the characteristics of the printed object. In order to be able to configure each of the extrusion parameters, it is necessary to use a "3D slicer" software. The main role of this program is to slice the 3D model into horizontal layers whose characteristics are determined by the chosen extrusion parameters. The next step the program performs is to calculate the amount of extruded material and the time it takes to perform the deposition operation, and, in the end, to transpose all this information into a G-Code file. In the end, the file is copied on an SD card memory and loaded onto machine. This, in turn, performs the instructions in the file just like any CNC machine.

For this article, the same two software will be used as in the case of setting up the parameters of the 3D printer. In order to compare, the same object will be sliced with the help of Z-Suite and an Open-Source software, 3D slicer. As documented below, the Z-Suite software Z-Suite is a proprietary one, offering less advanced options for configuring the printing parameters.

3.1. Layer height
The height of the deposited layer is the thickness of each layer, being determined by the length of the step of the extruder on the Z-axis before depositing each layer. The height of the deposited layer has the greatest influence on the resolution and time of obtaining the object. Depending on the desired characteristics of the printed object, a larger thickness of the deposited layer will be chosen if the appearance of the object is not a priority and a lower thickness otherwise. In the case of objects whose geometry is formed by inclined surfaces, it is favourable to obtain the object by depositing a larger number of layers resulting in a lower height of the deposited layer [10].

In the case of the printer used to make the support, the minimum height of the deposited layer is 150 μm, given the positioning accuracy of the extruder on the Z-axis. In addition, the minimum height of the deposited layer is also related to the nozzle diameter, i.e. 20% of it [11]. For a nozzle diameter of 0.5 mm, the recommended minimum height of the deposition layer is 0.1 mm. As a rule, a maximum height of the deposited layer is recommended with a value between 50% and 75% of the size of the nozzle. Although objects can be obtained using a layer height greater than the recommended layer, the effect will be a decrease in the mechanical strength of the object and an increase in the roughness of its surfaces [11].

Considering the above, within this article, given that the main purpose of the laptop / tablet support is to provide support, a height of the 0.375 mm deposition layer should be chosen, this being 75 % of the deposition head diameter value. This will result in a decrease in the time required to obtain the object in exchange for outer surfaces with a higher roughness. If the appearance is more important than other aspects, then the suggested value for layer height is 0.15 mm.

3.2. Layer width
The width of the deposited layer is determined by the feeding speed of the filament, the speed at which the extruder moves while depositing the filament and the diameter of the hot end diameter. The minimum width is determined by the diameter of the hot end, since during deposition the deposited

material needs to be pressed between the hot end and the previously deposited layer to adhere thereto. Thus, in order not to affect the integrity of the printed object, a minimum width of at least 20% greater than the diameter of the hot end is recommended [12]. The upper limit of the layer width is limited by the arcing phenomenon of the layer deposited under the influence of the pressure exerted by the hot end. It is recommended to use a width up to twice the size of the deposition nozzle diameter to avoid this phenomenon [12].

The optimal width of the deposited layer depends on the characteristics that the 3D printed object must possess and the ratio between the deposited layer height and the diameter of the hot end. In addition, the optimal width differs depending on the type of layer that is deposited (top, outer, or bottom). In depositing the bottom layer, the contact surface with the deposited deposition tray increases directly in proportion to its width, thus facilitating good adhesion.

Another important factor to consider when setting the layer width is the bond between the lateral surfaces of the filaments composing the deposited layers. This bond determines the roughness of the top layer and the inclined surfaces, but also the mechanical strength of the object under the action of a force applied vertically thereto, in other words a vertical shear force.

Only the Slic3r software allows the adjustment of the size of the connecting surface between the lateral surfaces of the deposited layers by offering the possibility to configure the overlapping coefficient between them. The overlapping coefficient represents the percentage of the deposited filament width dimension that overlaps with the deposited filament on one of its sides. It is necessary to keep in mind that in order to avoid a high roughness of the horizontal surfaces it is required that the volume of the free space between the deposited filaments is not exceeded.

For the support presented, a value of 1 mm of depositing layer and 15% of overlapping coefficient will be used. By choosing this value, good adhesion will be obtained between the lateral surfaces of the filaments deposited.

3.3. *Infill Percentage*

In the case of 3D filament printing technology, the term "infill" refers to the density of which the internal structure of the object is composed. This parameter influences the weight of the object, the required production time, the amount of material and the mechanical strength of the object. The higher the infill percentage is, the more resistant the object becomes and the deposition time, the amount of material required and the weight increase.

The degree of infill is measured as a percentage, so a 0% infill rate describes an object that does not have the inner structure and a 100% infill rate is an object whose internal structure is entirely composed of the material. According to [13], in most cases, 5% - 15% infill is enough and 100% infill is very rarely used. If an object is produced for appearance only it can be obtained using a low infill rate, thus significantly reducing the cost and time it takes to obtain it while an object whose role is to take over forces (e.g. laptop / tablet support) will need a higher degree of infill.

Taking into account that the laptop / tablet support will be subjected to compressive force it is necessary to consider the required resistance of the object so that it can exercise its role. According to the studies carried out [14], besides the infill degree, the mechanical strength of the objects in the case of compression is directly influenced by the height of the deposited layer. Due to the orientation chosen on the tray, the area of the laptop / tablet support that is in contact with it will support all the compressive force. To determine an optimal degree of infill, it is necessary to calculate the area and determine the pressure to be exerted on it. The program used to configure 3D print parameters does not allow measurement of this area, but this value can be extracted from CAD model. The cumulated area of the contact surface that are placed horizontally on the table is about 52 cm^2. The average weight of a laptop is about 2.5 kg, tablets being lighter that notebooks. Using the data from [14], the base surface of the support requires a minimum resistance of about 0.005 MPa, the resistance achieved for a linear fill model even using the worst-case parameters. Thus, in order to reduce the amount of material used and the time required to obtain the object, a filling degree of 10% will be used for the support. This decision is justified by the need for support for the upper layers of inclined surfaces present in the geometry of the object.

IOP Conf. Series: Materials Science and Engineering **682** (2019) 012015 doi:10.1088/1757-899X/682/1/012015

3.4. Infill pattern

The deposition pattern is the model inside the manufactured object, thus giving it a certain structure. There are a variety of available infill patterns, each of which has certain advantages and disadvantages. Depending on the geometry of the object, in addition to providing an internal structure of the object's resistance, the infill pattern has in some cases also a supporting role for the deposited layers, thus preventing the deformation of the object during the deposition if it has edges suspended in its geometry. The same pattern can be deposited using different filling degrees.

The most used models where the object must withstand compressive forces are linear, triangular and hexagonal [15]. The linear model is the standard infill model for objects obtained through the filament printing technology. It has resistance to the compressive forces applied to each of the three axes, requiring the shortest production time due to the reduced complexity of the extruder path during its deposition. The triangular pattern has an advantage if the outer walls of the object must withstand mechanical forces, giving the object an increased stiffness. This model requires a longer lead-time than the linear model. The hexagonal model is also popular, showing resistance to the compressive forces applied to each of the three axes. Based on the same study already mentioned [14], for the same degree of infill, the linear deposition pattern of the inner layers exhibits the highest mechanical strength in the case of compression, as presented in figure 2. In conclusion, for the support will be selected a linear pattern.

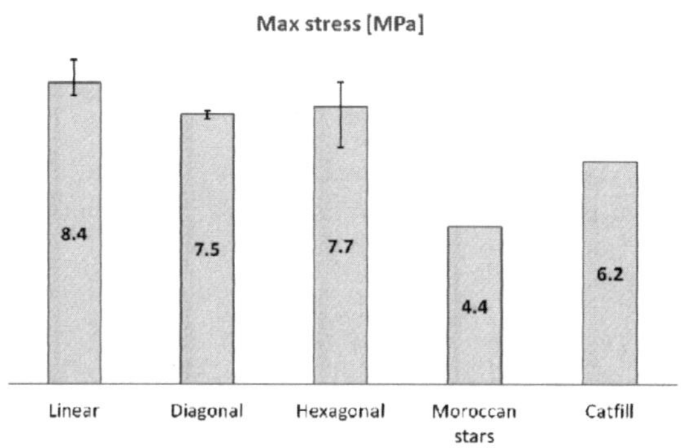

Figure 2. The relation between infill pattern and strength [14].

3.5. Number of layers

The outer layers can be viewed as a casing of the object. In the case of the additive filament printing technology, the outer layers are the first to be deposited on each layer [15]. The outer layer deposition path is defined by the outer contour of the 3D printed object.

The most important considerations to be taken in account for the outer layers are the mechanical strength, the possibility of finishing, the quality of the inclined surfaces, the time and the material required for the production, and the relationship between the diameter of the hot end and the dimensions of the outer layers. A higher number of vertical shells results in a thicker case of the object, and thus in an increased resistance of its mechanical strength. In this way, an increase in the robustness of the object is achieved without the need for an increase in the degree of filling. As with infill, the time and material requirements increase directly in proportion to the number of outer layers deposited.

Considering that the outer layers must provide mechanical resistance to the object, it is recommended to print them at a width twice the diameter of the print nozzle. This provides the required pressure between the deposited layer and the previously deposited layer thus facilitating good adhesion between the two layers. Because the mechanical strength of the object is already guaranteed by the fact that it has an infilling structure, the number of layers chosen is justified by obtaining the inclined surfaces without the visible filling structure of the object. According to the simulations performed in the Slic3r software, the number of two outer layers is enough to cover the object filling structure on the sloping surface area.

3.6. Travel speed

The travel speed is the speed of the extruder when it does not deposit material. The main benefit of a high travel speed is the decrease in the time that the object is printed. A high travel speed contributes to improving the appearance by eliminating the phenomenon of stringing of the filament during the repositioning of the extruder. The higher the repositioning distance is, the more the phenomenon of stringing is evident. Thus, by using a high repositioning velocity, the material present in the hot end is less likely to drain during repositioning. Since the support has, in the same layer, separate surfaces between them, it is necessary to use a higher travel speed to avoid the phenomenon of stringing of the filament. Theoretically, for the printer used, the maximum travel speed may reach 3.5 m/s if the repositioning distance allows the extruder to accelerate to that speed. However, a high acceleration of the extruder and a sudden stop when it has reached its destination may produce deviations from positioning accuracy due to inertia. Consequently, the maximum recommended travel speed is 250 mm/s.

3.7. Print speed

The print speed is the speed at which the deposition head (the extruder) moves while depositing the material. This speed is not limited by the mechanical structure of the printers or electric motors but by the amount of material the extruder can deposit in a unit of time. The main advantage of a high print speed is the decrease in the time required to obtain the object.

Consideration should also be given to the negative effects that a high print speed can have on a deposited object. The filament is subjected to expansion during deposition due to the high temperatures at which it is heated to reach the melting point. After being deposited, the plastic layer contracts while returning to ambient temperature. A small thickness of the deposited layer favours its rapid cooling, and because it is composed of a small amount of material, during contraction it is less prone to the arcing phenomenon. In other words, in order to avoid the arcing phenomenon of the deposited layers, they need a short cooling time before the next layer is deposited over them. Thus, the optimal deposition rate is a parameter that varies according to the horizontal plane dimensions of the object. When depositing exterior layers, it is advisable to use a lower deposition rate than the layers that make up the infilling structure because the aesthetic defects occurring in these layers will be visible.

Given that the object chosen in this project is of large size and the geometry of its layers is composed predominantly of straight lines, the probability of occurrence of the arcing phenomenon of the layers is relatively small. It is thus possible to choose a printing speed in order to decrease the required production time. Consequently, the selected print speed will be 80 mm/s.

4. Discussion and evaluation

The first step in obtaining an object through additive technology is to define its purpose and its features. Depending on this, 3D object modeling is done and a suitable printer is selected. After the object geometry is defined, the best positioning on the printing bed is identified in order to minimize defects and print time.

Specifically, for the analysed object, a PLA type material and a printer (Zortrax M200) with appropriate specifications are selected. The next step involves configuring print parameters. Choosing optimal parameters must be done in relation to the final characteristics of the object. In this article, the mechanical strength, the time to print and the amount of material used were considered. In Table 1, the influence of these parameters is presented, with a rating of 0 to 2, depending on the importance. If a criterion is not influenced by a parameter, it will get the 0 rating.

The optimal setting of the depositing parameters will consider the degree of influence of the print parameters according to those presented in Table 2. With Slic3r software can be estimated, deposition time and the quantity of material used. It is important to note that if the minimum quantity of material is to be used then the overlap coefficient is 0%, otherwise is 15%. The values for 3D printer parameters remain constant even in the optimal configuration because they are determined by the printer's technical specifications and the material used. The values for 3D printer parameters remain constant even in the optimal configuration because they are determined by the printer's technical specifications and the material used.

IOP Conf. Series: Materials Science and Engineering **682** (2019) 012015 doi:10.1088/1757-899X/682/1/012015

Table 1. 3D print parameters evaluation according to Slicer software.

	Parameter evaluation according to:			
	Appearance	Mechanical resistance	Time	The amount of material used
3D printer parameters				
Hot end diameter [mm]	1	0	1	0
Filament diameter [mm]	0	0	1	0
Extrusion temperature [°C]	1	1	0	0
3D print parameters				
Layer height [mm]	2	1	2	0
Layer width [mm]	1	2	0	1
Infill degree [%]	1	2	2	2
Infill pattern	0	2	2	0
Number of lower layers	1	1	1	1
Number of outer layers	2	1	1	1
Number of upper layers	2	1	1	1
Support structure	1	0	1	2
Travel speed [mm/s]	1	0	2	0
Printing speed [mm/s]	2	1	2	0

Table 2. Time and material specifications according to Slic3r software.

	Time and material criteria			
	Appearance	Mechanical resistance	Time	The volume of used material
3D printer parameters				
Hot end diameter [mm]	0.50			
Filament diameter [mm]	1.75			
Extrusion temperature [°C]	200			
3D print parameters				
Layer height [mm]	0.1	0.375	0.375	0.375
Layer width [mm]	0.6	1	1	0.6
Infill degree [%]	10	100	0	0
Infill pattern	Linear	Linear	Linear	Linear
Number of lower layers	3	3	1	1
Number of outer layers	3	2	1	1
Number of upper layers	3	2	1	1
Travel speed [mm/s]	250	250	250	250
Printing speed [mm/s]	50	80	100	100
Time and material requirements				
Time required [h]	14.90	5.50	1.20	1.50
Material required [cm^3]	87.20	202.40	46.40	29.60

5. Conclusions

In order to achieve the functions for the support presented, a linear pattern of the infill structure has been selected because it is suitably resisting the compressive forces to which the support is subjected. The

IOP Conf. Series: Materials Science and Engineering **682** (2019) 012015 doi:10.1088/1757-899X/682/1/012015

number of lower layers was three to give a solid foundation to the object. A value of two superior layers has been chosen to cover the infill pattern.

The height of 0.375 mm and the width of the outer layer of 1 mm are favourable for adhesion between the layers and to the printing bed. It was preferred the option of depositing two outer layers for covering the inclined surfaces without the filling structure of the object being visible. The deposition height of 0.375 mm represents 75% of the size of the deposition head, being the maximum value for a 0.5 mm diameter nozzle. This value generates less deposition time and less roughness. The 1 mm width of the depositing layer and a 15% interference coefficient gives a good adhesion between deposited layers.

The fill rate of 10% reduces the amount of material used and the printing time. In the same time, the printing speed of 80 mm/s significantly reduces the time required for printing. Since the support has larger dimensions and the geometry of the layers is composed predominantly of straight lines, the probability of arching layers phenomenon to occur is relatively small.

The purpose of the article was not to compare side by side the two Slicer programs. The reason why the two programs were discussed was the desire to control as many parameters as possible so that the final product corresponds best with the initial requirements. For the mentioned purpose, Slic3r is the variant that allows the most modifications. This feature may, however, be a drawback of proper use by a less experienced user. However, as the data in table 2 highlight, the variations, both in the quantities used and in relation to the manufacturing time, are significant. Nevertheless, when these issues are less important, then the Z-suite software is more suited to the average user. Technological aspects are either suggested or chosen by the program in the Automatic version. This facilitates the immediate obtaining of products through additive technology. Usually, commercially available 3D printers are delivered with own software. For replicated printers is necessary to use one of the available software. Thus, this article refers to this situation and when the user wants to optimize exploitation and filament consumption.

6. References

[1] Messimer S L, Patterson A E, Muna N and Deshpande A P 2018 Characterization and Processing Behavior of Heated Aluminum-Polycarbonate Composite Build Plates for the FDM Additive Manufacturing Process *J. Manuf. Mater. Process.* **12** 2-27

[2] Turner B N, Strong R and Gold S A 2014 A review of melt extrusion additive manufacturing processes: I. Process design and modelling *Rapid Prototyp. J.* **20** 3 192-204

[3] Information on https://zortrax.com/software/, accessed: 10.02.2019

[4] Information on https://slic3r.org/, accessed: 10.02.2019

[5] Information on https://store.zortrax.com/materials/zortrax-m200-z-pla-pro, accessed: 10.02.2019

[6] Vasilescu M D and Groza I V 2017 Influence of technological parameters on the roughness and dimension of flat parts generated by FDM 3D printing *Nonconventional Tech. Rev.* **3** 20-24

[7] Information on www.matterhackers.com/news/choosing-the-right-3d-print-bed-surface, accessed: 10.12.2018

[8] Information on https://zortrax.com/3d-printers/m200/, accessed: 10.12.2018

[9] Information on https://www.printitindustries.com/pages/print-speed-calculator, accessed: 10.12.2018

[10] Information on http://support.3dverkstan.se/article/30-getting-better-prints, accessed: 10.12.2018

[11] Information on https://rigid.ink/blogs/news/what-3d-printer-nozzle-size-should-i-use-the-pros-and-cons, accessed: 10.12.2018

[12] Information on http://manual.slic3r.org/advanced/flow-math, accessed: 10.12.2018

[13] Information on https://3dplatform.com/3d-printing-tech-tips-infill-percentage-and-pattern-explained/, accessed: 10.12.2018

[14] Information on What is the influence of infill, layer height and infill pattern on my 3D prints?, available at http://my3dmatter.com/influence-infill-layer-height-pattern/, accessed: 10.12.2018

[15] Information on https://www.3dhubs.com/knowledge-base/selecting-optimal-shell-and-infill-parameters-fdm-3d-printing, accessed: 10.12.2018

Experimental researches concerning the manufacture of multitool 3D printer, type DIY (do it yourself)

D I Băilă, S Tonoiu, M C Ștefan and I Sapadin

University Politehnica of Bucharest, Blv. Splaiul Independentei, no. 313, sector 6, cod 060042, Bucharest, Romania

baila_d@yahoo.com

Abstract. In the last decade, additive manufacturing has greatly developed and the manufacture of cheaper 3D hybrid (multitool) printers has increased. Hybrid 3D printers are equipped with interchangeable heads for manufacture and permit multiple types of processing such as 3D printing, CNC cutting, laser engraving or deposition of pasty materials. The objective of this paper was to manufacture a hybrid 3D printer prototype, type DIY (do it yourself) and was equipped with g-code software necessary for 3D printing process and for CNC cutting. The materials used for construction for this hybrid 3D printer are cheap and accessible and some of electronic components are reused and recovered from old equipment and the functional 3D hybrid printer obtained can be a model for sustainability of products.

1. Introduction

The purpose of this research was to create a hybrid 3D printer prototype, recycling and recovery of old electronic components, such as were used stepper motors from some old Xerox equipment, enabling sustainable development of the product. Hybrid 3D printers are multifunction printers that will replace traditional 3D printers in the future or min-CNC and will be finding in most households, because of their varied functionality. A hybrid 3D printer ZMorph is presented in figure 1 [1].

Hybrid 3D printers are equipped with several interchangeable heads: as simple 3D print head, 3D dual head, CNC head, laser engraving head and ceramic material deposition head.

Fused Deposition Modeling technology (FDM) uses a variety of filament-like materials of PLA, ABS, nylon, etc. [2] The mechanical strength of the manufactured parts is very good, often existing 3D printers that are sold with 3D printed components in their composition [3]. Hybrid 3D printers used this technology for printing 3D parts.

The mechanical strength of the 3D printed parts is very good, often existing 3D printers that are sold with 3D printed components in their composition [4].

The hybrid prototype 3D printer made in this research also has in its composition, 3D printed parts, 3D printed directly on itself, like the red brackets of the printed table and the head CNC components.

FDM technology is based on the materialization of a CAD product by adding successive layers. The object is saved in the stl. file to be used by 3D printer software. This technology permits the supports building necessary for the 3D printed parts [2, 3].

The prototypes obtained by FDM technology, do not require any additional post-processing treatments and can be used immediately, presenting a high quality surface [5].

The Additive Manufacturing technology helps identify any problems that may arise in the design and conception [6, 7].

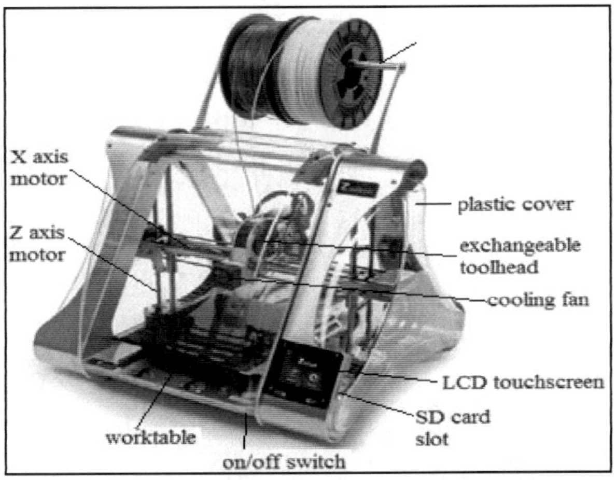

Figure 1. Commercial 3D hybrid printer ZMorph. [1]

ZMorph printer can print layers up to 50 microns. ZMorph 3D hybrid printer is equipped with five interchangeable heads allowing printing, milling, drilling and engraving materials widely. The manufacturing dimensions of ZMorph printer are 300x235x165 mm. The manufactured materials of this printer are very different from PLA, ABS, special filaments wood, etc. The software used of this printer is Voxelizer. The price of a hybrid 3D printer is quite high, around 4000 euros. This research tries to realize a low cost hybrid 3D printer, around 500 euros, that will be using for didactic and researches experiments [8, 9].

2. Experimental researches

The experimental research consists to manufacture a hybrid 3D printers DIY (do it yourself), low cost, using readily available cheap materials and tools.

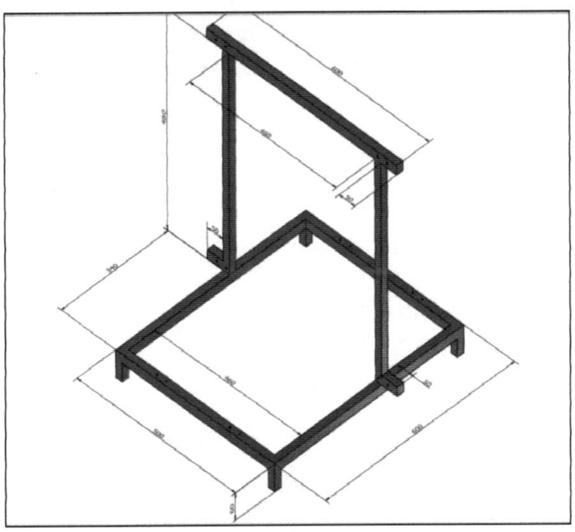

Figure 2. Structure design of 3D printer.

Figure 3. Assembly and welding of 3D printer parts.

In the design stage, it was considered the sizing of structural elements depending on the size of standard components used (clamps catching SK10, SK16, bearings SC10UU, linear guide Ø10, motors NEMA17) in construction of 3D printer structure, which will be manufactured in welded rectangular profile 20x20 mm, as follows in figure 2 below.

In figure 3, it was realized the cutting and drilling plates of PTFE material which will form the printing table, it has been the attaching of the stirrups and linear guides, to the fixing table and uncoupling the bed springs and adjustable thumbscrew to facilitate leveling.
It made cutting boards and were positioned linear bearings. NEMA17 motors were attached and completed the construction.

Concerning the electrical connections, it has realized different connections between controller and stepper motors. In order to achieve electrical control network used: a development board - Arduino MEGA 2560, a 1.4 Ramps SHIELD module, 5 Drivers 4988, as in Fig.4.

Figure 4. Arduino MEGA 2560 board.

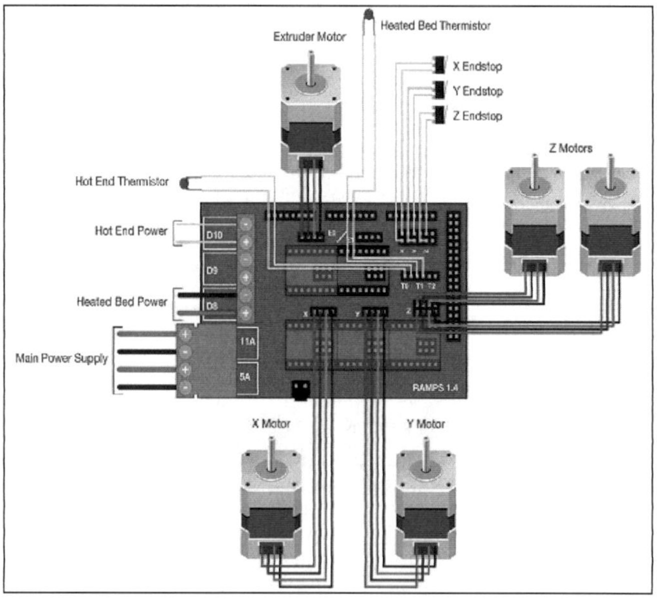

Figure 5. Electrical installation scheme.

In figure 5 is presented the electrical installation scheme. It used a power supply PC PSU 12V 14.6 and an endstop with mechanical feeler as in figure 6 and figure 7.

Figure 6. Power supply PC PSU 12V 14.6.

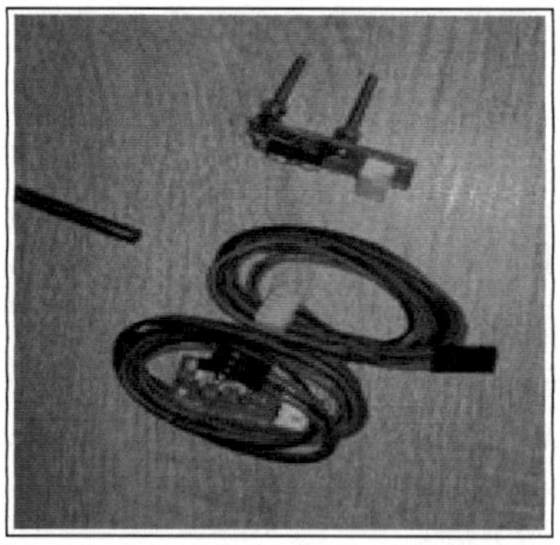

Figure 7. Endstop with mechanical feeler.

The single-pole stepping motors with 4 phases have been converted into 2 phases by eliminating " the mid-point" of the coils A and B (MA:MB), as in figure 8.

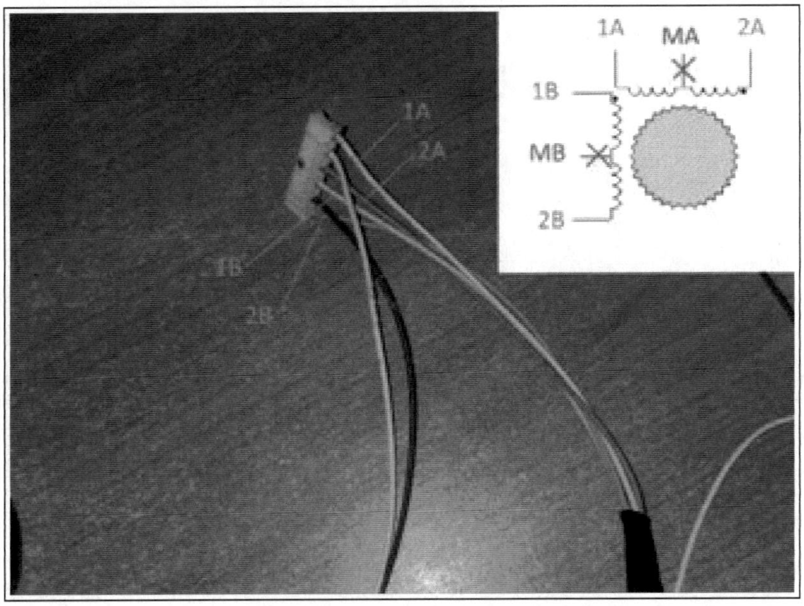

Figure 8. Stepping motors.

For 3D printer operation was performed scheduling and optimization software. Implementation of physical parameters of the printer was done using Marlin1.1 code as in figure 9, uploaded on the development board Arduino MEGA 2560.

```
/**
 * Default Axis Steps Per Unit (steps/mm)
 * Override with M92
 *                                      X, Y, Z, E0 [, E1[, E2[, E3[, E4]]]]
 */
#define DEFAULT_AXIS_STEPS_PER_UNIT   { 80, 100, 1600, 94.3 }
                                  Fulie z20   Fulie z16   Surub P2   Extruder MK8
/**
 * Default Max Feed Rate (mm/s)
 * Override with M203
 *                                      X, Y, Z, E0 [, E1[, E2[, E3[, E4]]]]
 */
#define DEFAULT_MAX_FEEDRATE          { 300, 300, 5, 25 }

/**
 * Default Max Acceleration (change/s) change = mm/s
 * (Maximum start speed for accelerated moves)
 * Override with M201
 *                                      X, Y, Z, E0 [, E1[, E2[, E3[, E4]]]]
 */
#define DEFAULT_MAX_ACCELERATION      { 2500, 2500, 2500, 10000 }

/**
 * Default Acceleration (change/s) change = mm/s
 * Override with M204
```

Figure 9. Marlin1.1 code used for 3D hybrid printer prototype.

IOP Conf. Series: Materials Science and Engineering **682** (2019) 012016 doi:10.1088/1757-899X/682/1/012016

```
#define TEMP_SENSOR_0 1
#define TEMP_SENSOR_1 0
#define TEMP_SENSOR_2 0
#define TEMP_SENSOR_3 0
#define TEMP_SENSOR_4 0
#define TEMP_SENSOR_BED 1
#define TEMP_SENSOR_CHAMBER 0
```

Figure 10. Thermistor type used (100K).

With its aid were set issues, like in figure 10:
- Port for the connection;
- Extruders number;
- Existence heated bed;
-Thermistor type used (100K);
-Maximum axes travel;
-Required number of steps per mm (the pitch of the threads);
-Point0 (homing point);
-Size table printed, as in figure 11;
-Maximum acceleration, as in figure 12;
-Maximum voltage motors.

```
#define X_HOME_DIR -1
#define Y_HOME_DIR -1
#define Z_HOME_DIR -1

// @section machine

// The size of the print bed
#define X_BED_SIZE 260
#define Y_BED_SIZE 260
// Travel limits (mm) after homing, corresponding to endstop positions.
#define X_MIN_POS 0
#define Y_MIN_POS 0
#define Z_MIN_POS 0
#define X_MAX_POS 260
#define Y_MAX_POS 260
#define Z_MAX_POS 280
```

Figure 11. Size table printed programming.

```
#define DEFAULT_ACCELERATION          2500    // X, Y, Z and E acceleration for printing moves
#define DEFAULT_RETRACT_ACCELERATION  3000    // E acceleration for retracts
#define DEFAULT_TRAVEL_ACCELERATION   2500    // X, Y, Z acceleration for travel (non printing) moves
```

Figure 12. Maximum acceleration programming.

G code can be obtained using various programs using models in a "Stl" as: Slic3r and its extension Pronterface or Simplify 3D, both programs also providing a good calibration printing parameters (printing speed, infill, thickness the outer walls, the height of the layers, the width of the filament, temperature) shown in figure 13. In figure 14 are presented the physical 3D parts obtained by 3D printing using the 3D hybrid printer prototype.

Figure 13. Simplify 3D program used to create g-code's.

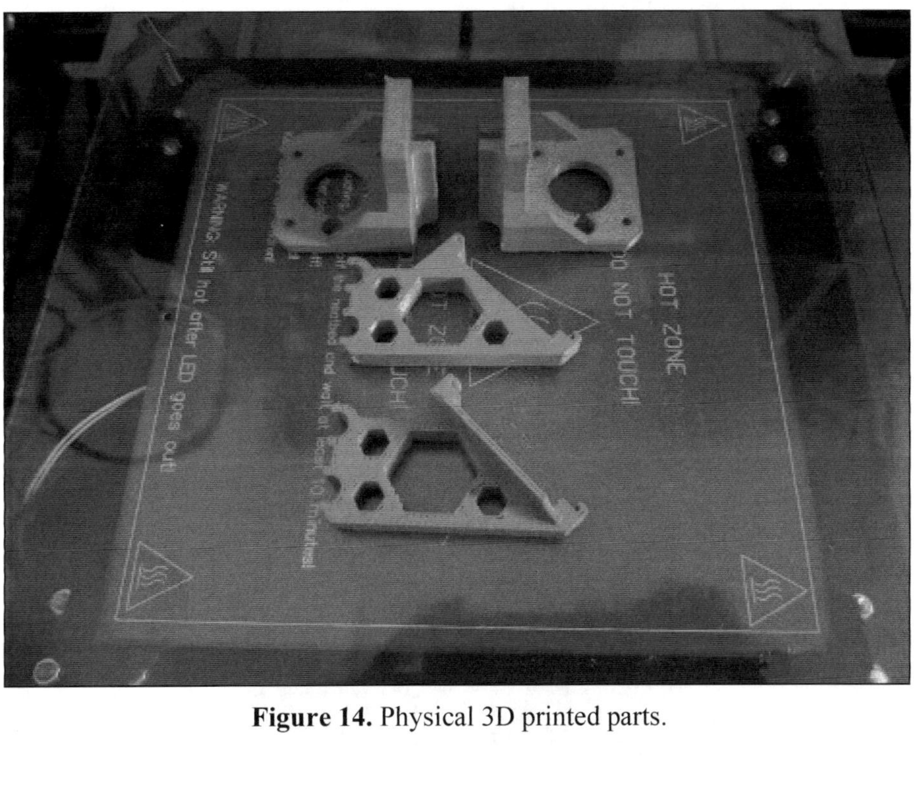

Figure 14. Physical 3D printed parts.

In figure 15 are shown the components of the hybrid 3D printer. In figure 16 are presented the exchangeable CNC tool head, some components of them are manufacture by 3D printing. In figure 17 is presented the CNC drilling on the 3D hybrid printer, using the CNC head.

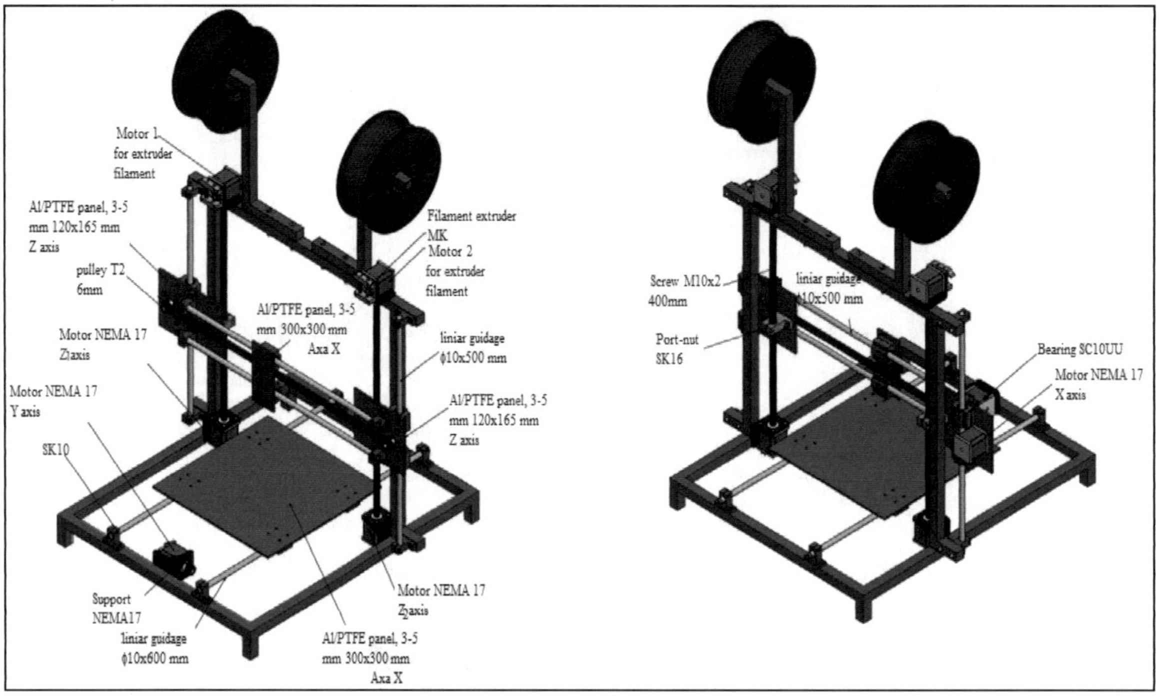

Figure 15. Components of hybrid 3D printer.

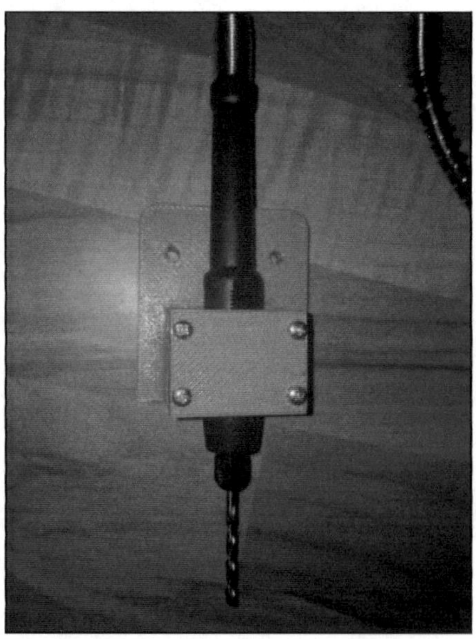

Figure 16. CNC head of 3D hybrid printer.

Figure 17. CNC drilling on the 3D hybrid printer.

3. Conclusions

This article presented the practical realization of a hybrid 3D printers DIY (do it yourself) using existing tools and inexpensive materials in any household.

Some of the elements used in hybrid 3D printer were recycled from old electronic equipment; this may lead to the sustainable development of the product.

Hybrid prototype 3D printer made in this research has been done on a low budget and is equipped with simple 3D print head and CNC head. The software used to perform the G-code has been Simplify 3D. The innovative part of this research was the electronics, the programming and optimization software. The dimensions of pieces manufactured by this 3D hybrid printer are 300x300x200 mm, having a very well precision of 0.2 mm.

4. References

[1] Information on https://zmorph3d.com/frontpage

[2] Berce P et al 2015 *Medical applications of Additive Manufacturing technologies* (Bucharest: Romanian Academy Publishing House)

[3] Berce P, Bâlc N, Ancău M, et al 2000 *Rapid Prototyping Manufacturing Technologies* (Bucharest: Technical Publishing House)

[4] Băilă D 2016 *Tehnologii si echipamente de Fabricare aditiva cu laser (ALM) a pulberilor metalice* (Bucharest: Ed. Printech)

[5] Băilă D 2016 *Biomateriale utilizate in Additive Manufacturing* (Bucharest: Ed. Printech)

[6] Rontescu C, Chivu O, Pacioaga A, Iacobescu G and Vasile I 2017 Research on Biocompatible Titanium Alloys Test Samples Obtained by Sintering *Rev. Chimie* **6** (67) 1124-1127

[7] Păcurar R, Bâlc N and Prem F 2011 Research on how to improve the accuracy of the SLM metallic parts *AIP Conference Proceedings* **1353**(1) 1385-1390

[8] Information on https://www.instructables.com/id/A-low-cost-3D-printer-with-basic-tools/

[9] Marlin Firmware- http://marlinfw.org/

IOP Conf. Series: Materials Science and Engineering **682** (2019) 012017 doi:10.1088/1757-899X/682/1/012017

Lean Business Practice as Support to QMS Implementation in According to ISO 9001:2015

M Djapic[1], P Popovic[2] and L Lukic[1]

[1]University of Kragujevac, Faculty of Mech. & Civil Engineering, Dositejava 19, Kraljevo, Serbia
[2]University of Belgrade, Institute VINCA, Mike Petrovica 12-14, Vinca, Serbia

djapic.m@mfkv.kg.ac.rs

Abstract. The contemporary market conditions impose increasingly strict requirements on organizations. Orientation towards product quality is one of the optimum strategies enabling survival and development in the ever more globalized market. The business practice required by ISO 9001 standard is often insufficient for an organization to survive in the market. Therefore many organizations, in addition to developing and implementing Quality Management System (QMS), optimize their operative processes by implementing the Lean tools. The paper provides a concept of interconnecting the ISO 9001:2015 standard requirements with the possible Lean tools deployment, which is to contribute to promoting effectiveness of the implemented QMS.

1. Introduction

The growth in productivity has nowadays brought about significant disproportion between the offer and demand of most industrial products. The manufacturing capacities installed are capable of producing much more products than the present day population needs. When this is joined with the achieved transport and communication systems improvements, we arrive to the basic reason for the developed world's decision to extend the free trade zone, which is named globalization today.

New and increasingly strict requirements have been imposed on manufacturing organizations (figure 1). On the one hand, these requirements are mirrored in more fierce competition, while on the other hand there have been continuous changes in customers' demands. For an organization to be competitive, it has to deliver products which fully meet customer's requirements, and it has to do this in the shortest time period and with the lowest costs related to the product's life. Survival of an organization is related to its capability to transform itself and fulfill these conditions. Those organizations which do not adapt to these changes and which do not find the way to satisfy customers do not have a chance to endure.

The decision to achieve satisfaction of customers through quality of products or services represents a major strategic change in many organizations. In that way, quality becomes the strategic tool by which competitive position is achieved, i.e. improving and maintaining quality becomes an important segment of the business strategy of any organization which wants to be the market leader.

Today, in product quality achievement, organizations typically rely on the international standards of the ISO 9000 series [1-3], i.e. on the requirements of the ISO 9001:2015 standard [1].

Is the business practice [4] imposed by the ISO 9000 series standards sufficient to respond to the contemporary market's requirements? It seems that it is not [5]. For that reason, many organizations striving to reach the best business practice, implement the Lean concept based on Toyota manufacturing system which has been developing since the beginning of the fifties of the last century until now [10-15].

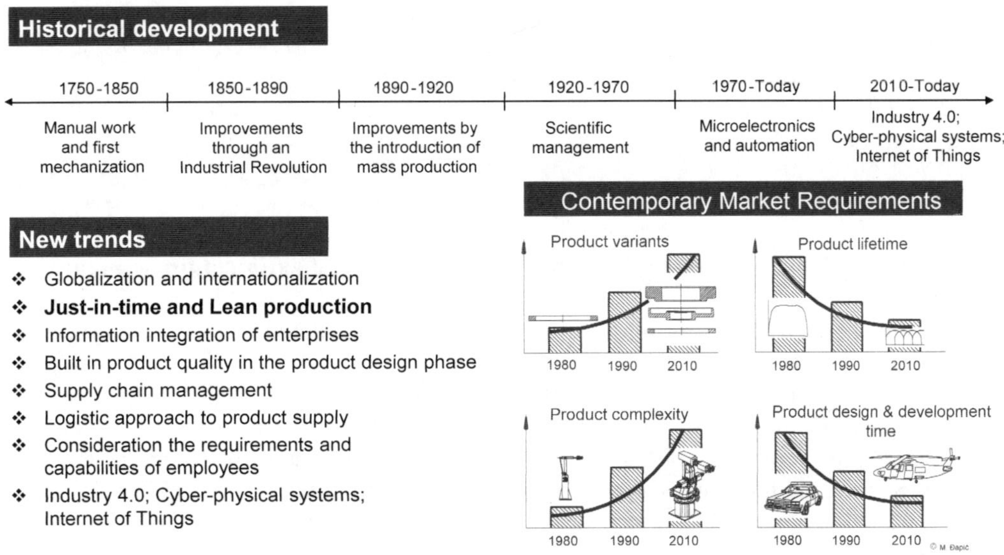

Figure 1. Manufacturing development trends

This paper provides the concept of linking the Lean tools, ISO 9001:2015 standard requirements, PDCA planning cycles and levels, which certainly can contribute to improving the QMS implementation in organizations. With this aim in mind, first the answer is given to the question: What is the best business practice and how to define it? On the basis of that, a business practice is presented which is advocated by ISO 9001:2015 standard and by Lean concept. In the end, the concept (framework) is given on the basis of which integration is possible of Lean tools into the procedure of QMS implementation.

2. What is the best practice and how to define it?

The best practice is the concept we come across in everyday life and at work. The Internet search enables finding thousands of books, papers, studies, etc., which contain in their headlines, among other things, the words "best practice". Few of these provide an explicit definition regarding what is understood by this concept. The reasons for this is probably that the authors believe it is the generally accepted concept understood by majority of people in the roughly the same way, so there is no need to define it specifically.

The similar situation is, for example, with the concept of "good engineering practice" appearing in the New Approach Directives within the EU technical legislation.

In order to reduce confusion and erroneous interpretations, it is necessary to have certain terms explicitly defined. As the confirmation of this, we shall use the Van Haren Publishing approach [4]. They have perceived that there is a great confusion in using the terms, and in understanding their meanings, such as in the case of "standard", "best practice", "body of knowledge", "framework", "guidance", "method", "model", etc. With the aim of promoting and establishing consistent usage of these terms in their publications, they have defined meanings of the above terms, based on extensive analyses of the best practice and thanks to contributions of a wide group of stakeholders.

To achieve this, it is best to first define what the "practice" means. Thus, "practice" is defined as [4]: "the way in which professionals work within their profession, in order to carry out a specific task". In that respect, "the best practice" can be defined as description of the best way of working based on the situation at hand".

On the other hand, one of the world's most renowned organizations in the field of quality, ASQ (American Society of Quality), defines "the best practice" [6] as: "A superior method or innovative practice that contributes to the improved performance of an organization, usually recognized as best by other peer organizations“. Two things are important here. First, in order to consider a business practice as the best one, it has to enable significant improvement of the organization’s performances. Second, such a practice has to be recognized by peer organizations as the best one.

The good practice is advocated, i.e. required, by the international ISO standards for management systems and is recognized by the widest grouping of stakeholders.

3. Best Business Practice and ISO 9001 Standard Requirements

The business management systems are established and implemented in order to enable:

- **<u>good practice</u> for conducting jobs and governing the organization at the management level** and to provide
- **consistent** (dependable, constant in the same manner, considered in the long run) **progress of activities.**

They contain multiple functions/disciplines. The implementation methods differ.

Thus, for example, Quality Management System (the management system for quality according to ISO 9001 standard requirements) – QMS, represents a part of an organization’s management system which is oriented towards creating new and maintaining existing customers, through the processes: understanding customers’ needs, planning, designing and delivering products and/or services which satisfy these needs.

As mentioned above, good practice is needed so that optimum output result can be provided, i.e. so that it has become the optimum method for ensuring these outputs. Good practice has in time been incorporated into the requirements of the standardized management systems.

According to [7], for instance, the good practice requirements have been incorporated into the ISO 9001 standard (figure 2), from the following three fields:

- Good practice for inspection of products/services and of processes outputs;
- Good practice for the top management activities aimed at supporting the good practice for inspection of products/services and process outputs.
- Good practice for transparency in tasks flow within organization

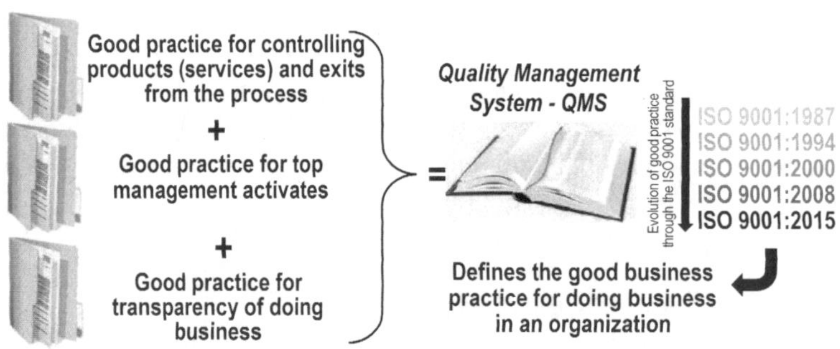

Figure 2. Good business practice incorporated into ISO 9001 standard (Adapted on the basis of [7])

Good practice is similar to a “movable target” as it changes and evolves continuously, which brings about a sort of management systems reviewing and improving every ten years or so. The management systems incorporate requirements for the good practice which was proven as valid in the previous period.

Consequently, if ISO 9001:2015 standard is observed and its changes with respect to the 2008 version, it can be stated that there are changes in the structure, terminology, requirements and in the elements that are more pronounced, but there are no changes in the purpose, objectives and scope of the standard implementation.

These changes in the requirements call for the organizations to implement about 18 % of new business practice requirements, to modify about 25% of the existing practices, to maintain the existing practice in 50% and about 7% of requirements impose withdrawal of the previously implemented business practices according to the 2008 version requirements [8]. In taking into consideration the changes in the standards structure, as well as in term definitions, and the new and modified requirements, it can be stated that transition from the 2008 version to 2015 version represents the most voluminous modification (revision) of ISO 9001 standard so far [9].

Now it is expected from organizations to implement that practice through development, implementation and certification of their management systems by which they secure carrying out of business processes in a modern and transparent way.

4. Lean Concept – As The Best Business Practice

The Lean manufacturing concept has been developed by the Japanese automotive industry. The initiative has been led by Toyota through the development of the Toyota Production System (TPS), in striving after the WWII to renew the Japanese industry. The Lean concept was presented to the rest of the world by the book by Womack, Jones, and Roos "The Machine That Changed the World" in 1990 [11].

There is today an extensive literature that covers all the aspects of Lean, whether explaining what it is or explicating how to implement the concept in actual conditions. We are presenting the Lean concept in the way as it is given in references from [10] to [15].

Lean signifies a business philosophy which advocates elimination of unnecessary losses in all manufacturing activities which are connected with: workforce, suppliers, technologies, materials and inventory management.

Basically, Lean stands for a systematic approach to identifying and removing wastes through continual improvements of these ways, so that individual objects manufactured (parts, assemblies or products) are moved in the manufacturing pace determined (i.e. "pulled") by customers' requirements (external /internal), in the striving towards perfection, in the business sense of the word [10].

Lean production stands for everything that is opposite to mass production. The manufacturing process is to "flow" continually, so that: flow of materials is as linear as possible, without turbulences, unnecessary straying, holds and backward paths. Many companies today try to implement Lean manufacturing abandoning the mass production concept, with the goal of: (1) quality improvement, (2) eliminating manufacturing wastes, (3) reducing products delivery times and (4) decreasing overall costs. Lean production puts the accent on preventing possible wastes like those of additional: times, manpower or used materials which are paid by customers, but which do not contribute to increasing the product value. Manufacturing is focused on the requirements of actual buyers and not to requirements of imaginary possible buyers which the organization is yet to locate. Hence, nothing is manufactured for the stock.

And finally: "*Lean production uses only half of everything compared to the mass production. It only uses half of the human effort in the factory, half of the manufacturing space, half of the investment in tools and half of the engineering hours to develop a new product in half the time. It also requires keeping half less inventory, which all together results in a significant decrease of non-harmonized products, i.e. scrap*" [11].

In order for an organization to survive (achieve success) in the global market, it has to:

- Be oriented towards customers, their wishes, requirements and expectations – since today, there is much higher number of products than customers who are ready to buy these;
- To react quickly towards fulfilling customers' requirements;
- To differ with regard to competition. Products of that organization have to be more innovative than those of competition and/or to differ regarding price, quality, etc.
- To continually improve its system.

All this can be achieved by fulfilling the Lean manufacturing goals, by application of Lean principles, i.e. by setting up and maintaining the Lean manufacturing systems. Now, having defined what is needed for an organization to be successful in contemporary conditions, let us see how Lean fits into these requirements.

The crucial idea behind Lean is to maximize the value for customer on the one hand and on the other hand to achieve this by minimizing wastes (holdbacks, scattering). In other words, the focus is on creating higher value for the customer by using: less resources, shorter time cycles, faster processing procedures and higher quality.

Lean initiative in any organization has four main goals:

1. Improvement of quality
2. Elimination of manufacturing wastes - scrap
3. Reducing the lead (main) manufacturing time
4. Reducing overall manufacturing costs.

If an organization is to reach the Lean goals, it has to implement into its business practice the set of Lean principles.

These principles are as follows:

- The Lean initiative begins so that the organization specifies values for its customers;
- Then it identifies the flow of values through the business-manufacturing processes which enable creating of value for the customer;
- Then, it has to ensure that the value creating activities flow continually;
- Manufacturing has to be "pulled" by the customer's requirements, and
- All the previous steps have to be continuously upgraded until reaching perfection in the business sense of the word.

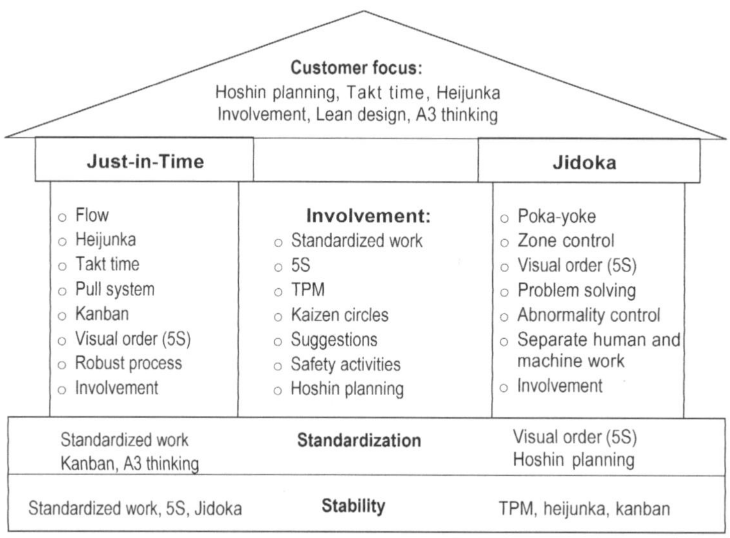

Figure 3. Toyota manufacturing system house (Adapted on the basis of [14])

Implementation of Lean tools and techniques (figure 3) on the operation level enables the company to satisfy their customers' demands regarding quality of products or services, the time when products need to be delivered and the price at which this can be done. The Lean manufacturing tools create business and manufacturing processes that are agile and efficient. Lean business practice aids companies in managing their overall costs while shareholders receive return on their investments.

In the end, it should be said that these tools have not originated as the result of an "ad-hoc" innovative capabilities of engineers and other employees working in Toyota, but as systematic solutions which have lead to the fulfillment of the global goal set by Toyota at the beginning of the fifties of the previous century. Because, as Taiichi Ohno once said, "All we are doing (in Toyota) is looking at the time line, from the moment the customer gives us an order to the moment when we collect the cash. And we are reducing the time line by reducing the non-value adding wastes. In order to reduce this time line, we had to eliminate all wastes, all scattering, i.e. all the activities which do not add value for the customer". This was the global initiator and the objective which has brought about systematic development and implementation of all these tools.

5. Lean Business Practice as Support to QMS Implementation

Lean business practice is nowadays considered the best business practice in the industry of individual material products, as covered by the term "manufacturing".

On the other hand, if implemented in the right way, ISO 9001:2015 standard requirements will certainly upgrade the organization, i.e. they will bring about implementation of the good business practice. If, for instance, operative processes in product realization are observed, as covered by the requirements of the Clause 8, it can be asserted that these requirements are insufficient for operative processes to respond to the modern market requirements. This can be obvious if the two groups of requirements are compared. The first set of requirements relates to the requirements defined in ISO 9001 standards, while the second group relates to the requirements put forward by the market itself, relating to the conditions of survival in it (Figure 1). This deficiency was perceived by many, and one of the first ones was Juran in the 2000 version of the standard [5].

Therefore, for an organization to respond to the modern market requirements, its operative processes have to be meaningfully upgraded, i.e. the best business practice is to be implemented as offered by the Lean methods and tools (Figure 3), together with the Six Sigma concept implementation. Lean process provides for cleaning the processes of the observed shortcomings, i.e. losses and wasting, while Six Sigma provides for stabilization of these.

Lean tools implementation enables cleaning of the business processes errors, where manufacturing is focused only to what customers require, where inventories only consist of items necessary for that particular production, where manufacturing is "pulled" in the rhythm determined by customer's requests, without bottlenecks or waiting. New "cellular" equipment arrangement is being introduced, with special attention directed to maintaining the equipment. The transition from one product series to the next has been optimized. Employees take meaningful participation in discovering and removing the observed errors. Implementation of Lean implies the process of continual improvements by organizing and implementing the "Rapid Improvement Event" – Kaizen) with participation of direct operators. On the other hand, as depicted in figure 4, QMS implementation according to ISO 9001:2015 requirements implies the deployment of PDCA cycles. It is therefore possible to identify clauses of the standard which are related to the segment "Plan" – Clauses 4, 5, 6 and 7, segment „Do" - Clause 8, segment "Check" - Clause 9 and segment "Act" - Clause 10.

The development and implementation of QMS also tackles the organization's activities in the strategic, tactical and operative levels.

In the strategic level, the organization's business operation is defined through defining the mission, vision and strategic directions of the organization's development in the period of three to five years. The business activities defined in this way are complemented by defining the authorizations and responsibilities, policies, principles and values that will be cherished by the organization's members, i.e. these are to be adhered to in the process of reaching the strategic goals defined in the vision. All the above stated is contained in the strategic document named "The strategic plan for organizational improvement /development". Related to the above stated, business processes are defined at the strategic level in the phase of QMS, together with their mutual connections in the procedure of delivering the products which have to satisfy customers and other interested organizations. In realizing these activities, it is very useful to apply the Lean VSM tool which relates to Value Stream Mapping.

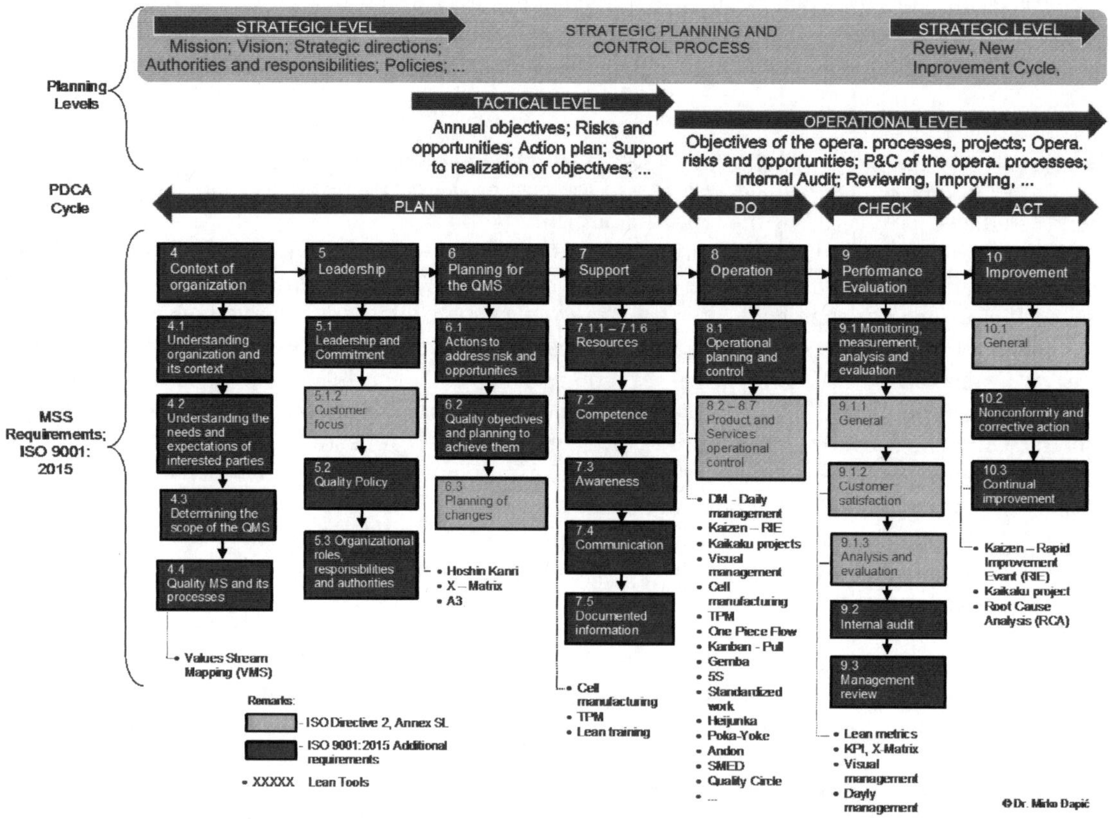

Figure 4. Linkage of the organization planning level, PDCA cycles methodology, ISO 9001:2015 standard requirements and possible implementations of Lean tools

From the Lean tools point of view, it is possible, at the tactical level, to deploy several tools. Among the first, there is the Hoshin Kanri one which ensures deployment of quality goals and strategies from top towards lower organizational units, i.e. towards manufacturing processes. It is also useful, in the procedure of defining resources for action plan realization, to deploy the cellular manufacturing concept, i.e. to install connected cellular manufacturing systems which will be maintained in accordance with the Total Productive Maintenance, (TPM) concept. As regards upgrading of the employees' awareness, knowledge and capabilities, it is possible to apply the "Quality Circle" concept.

The operative level involves planning and management of business/manufacturing (operative) processes execution. The risks related to realizing these processes' goals are assessed, and possible opportunities are identified for their upgrading. Monitoring of their performances is done, as well as defining of possibilities for upgrading.

The major portion of the developed and implemented Lean tools is applied, especially in the automotive industry, at the operative level. Consequently, in implementing the Clause 8 of ISO 9001:2015 standard, an entire set of Lean tools can be used, as shown in figure 4. The analysis of these tools deployment possibilities goes beyond the aims and scope of this paper, since a detailed analysis of the Lean tool possible applications in satisfying the requirements of clauses 8, 9 and 10 requires much more space.

6. Conclusion

Contemporary market conditions have been posing new and ever more severe requirements to organizations. In the situation where there are much more products than customers ready to buy them, many organizations discover an optimum strategy of survival and growth in their orientation towards product quality and also in the quality of the entire organization.

Today, the most widely spread approach to achieving product quality is through the process of developing and deploying the quality management system (QMS) according to the ISO 9001:2015 standard requirements. The business practice established in that way is insufficient for an organization to become a market leader. The orientation towards Lean & Six Sigma manufacturing philosophy is the very thing that has enabled many organizations to achieve that objective.

Following this trail, the paper analyses the business practice imposed by the new version of ISO 9001:2015 standard requirements and by the Lean manufacturing practice concept, i.e. analyses their compatibility. In realizing the operative processes, the implementation of ISO 9001:2015, analyses and improvements of these processes draw main benefits from implementing the Lean tools. For that reason, the operative level of Lean deployment is advocated, so that after cleaning the business-manufacturing processes from errors and after optimizing their performances, the obtained state is standardized and built in as the QMS constituent part. In that respect, many organizations, simultaneously with the QMS development and implementation, also optimize their operative business-manufacturing processes by deploying the Lean tools.

7. References

[1] ISO 2015 *ISO 9001:2015 - Quality Management System – Requirements* ISO

[2] ISO 2015 *ISO 9000:2015 - Quality management systems — Fundamentals and vocabulary* ISO

[3] BSI 2016 PD ISO/TS 9002:2016 - *Quality management systems — Guidelines for the application of ISO 9001:2015* The British Standards Institution

[4] Van Haren Publishing 2018 *Best practics, Model, Framework, Method, Guidance, and Standard: towards a consistent use terminology – revised* (Available on: www.vanharen.net. Accessed: 15.08.2019)

[5] Djapic M 2005 *Evidence systems in the development of products and processes* (Monograph no. 9 in the series of monograph papers "Intelligent Manufacturing Systems") (Belgrade: LOLA Institute)

[6] ASQ 2018 *Quality Glossary* (Available on: www.asq.org. Accessed: 28.08.2019)

[7] Ames M 2017 *Quality Management System: Background, Evolution, and Future of ISO 9001* (ASQ Webinar: Available on: www.asq.org, Accessed 25.07.2019)

[8] Hoyle D 2017 *ISO 9000 Quality System Handbook – Updated for the use ISO 9001:2015 standard* (New York: Routledge) p 893

[9] Abuhav I 2017 *ISO 9001:2015 – Complete Guide to Quality Management Systems* (Boca Raton: Taylor & Francis Group) p 444

[10] Macinnes L R 2002 *The Lean Enterprise Memory Jogger: Create Value and Eliminate Waste Throughout Your Company* (Salem: GOAL/QPC)

[11] Womack P J, Daniel T J and Macmillan T 1990 *The Machine that Change the World* (New York: Raw Associates and Toronto: Collier Macmillan) p 320

[12] Koenigsaecker G 2009 *Leading the Lean enterprise transformation* (Second Edition) (Boca Raton: Taylor & Francis Group) p 261

[13] Manos A and Vincent C 2012 *The Lean Handbook, A Guide to the Bronze Certification Body of Knowledge* (Milwaukee: ASQ Quality Press) p 544

[14] Dennis P 2015 *Lean Production Simplified: A Plan Language Guide to the World's Most Powerful Production System* (Third Edition) (Boca Raton: Taylor & Francis Group) p 243

[15] Monden Y 2012 *TOYOTA Production System, An Integrated Approach to Just-In-Time* (Fourth Edition) (Boca Raton: Taylor & Francis Group) p 548

Acknowledgment

This research was supported by Ministry of Education, Science and Technological Development of Republic of Serbia under Projects TR35031.

IOP Conf. Series: Materials Science and Engineering **682** (2019) 012018 doi:10.1088/1757-899X/682/1/012018

Process planning solution for increasing the productivity and robustness of work schedules for batch production processes

M G Catană, M R Dijmărescu and N Tunsoiu

University POLITEHNICA of Bucharest, Manufacturing Engineering Department, 313 Splaiul Independentei, 060042 Bucharest, Romania

mg_catana@yahoo.com, manuela.dijmarescu@imst.pub.ro, tunsoiunicolae@gmail.com

Abstract. This paper analyses the issue of ensuring enhanced productivity and robustness to work schedules designed for multi-operational batch production processes, when a fixed, limited number of workstations can be used for production. The productivity of a batch production schedule for a specific quantity of items is measured by the length of its cycle time, while the robustness of the schedule is given by the size and distribution of operational slack times that are ensured during the production process run. It is mathematically demonstrated that a simple process planning solution may be the key for meeting the above mentioned objectives, if it is possible to be applied in practice in combination with the lot streaming strategy for the production flow.

1. Introduction

Batch production is a manufacturing strategy used for manufacturing large quantities of items in periodical shares consisting in small-sizes, medium-sized or large-sized production batches [1, 2]. The units in each batch are moved from one workstation to another by using special transportation means. The entire production batch can be transferred at once in case of small-sized batches, or, in the cases of small-sized, medium-sized or large-sized batches, a fraction of production batch termed transfer batch is transported between the workstations of consecutive operations [1-3].

The two manufacturing flow strategies most used today for batch production are production batch flow strategy (BPF) and transfer batch flow strategy (BTF).

BPF is the manufacturing flow strategy that implies the transfer of the entire production batch from preceding operation with index number k to succeeding operation with index number $k+1$ [1-3]. The strategy is frequently used for small-sized batches.

BTF is the manufacturing flow strategy that implies the transfer of fractions of production batch known as transfer batches from operation k to operation $k+1$ [1-3]. This strategy may be used whatever the size of production batch (N), by adopting a size of transfer batch (N_t) smaller than that of production batch: $N_t < N$ [1-3]. BTF strategy is frequently termed lot streaming strategy [4-10].

In order to deliver the transfer batches from preceding operation k to succeeding operation $k+1$ so that no idle times or work-item shortages appear in operation $k+1$, a minimum start-to-start lag time is necessary to be ensured for operations in case of BTF strategy [1, 2, 4, 9]. Assuming that setup times of operations are negligible with respect to their run times, the minimum start-to-start lag time ($D_{k,k+1}$) of consecutive operations k and $k+1$ can be computed as follows [2]:

$$D_{k,k+1} = \begin{cases} N_t \cdot T_{uk}, & if\ T_{uk} \le T_{uk+1} \\ N_t \cdot T_{uk} - (N - N_t) \cdot T_{uk+1}, & if\ T_{uk} > T_{uk+1} \end{cases} \quad [\text{min}] \tag{1.1}$$

where: T_{uk} and T_{uk+1} are run times for manufacturing operations k and $k+1$, in min/unit; N_t – size of each transfer batch delivered between operations k and $k+1$; N – size of production batch.

The productivity of a batch production process that produces a quantity of N items can be measured by the length of its production cycle time (T_{cN}), so that the shorter the production cycle time, the higher the productivity of the process.

In the case of BPF strategy, minimum length of batch production cycle time can be calculated with equation (1.2), while for BTF strategy minimum length of batch production cycle time can be estimated with equation (1.3) [1, 2]. In both cases, setup times of operations are considered negligible with respect to operation run times.

$$T_{cN} = N \cdot \sum_{k=1}^{n} T_{uk}\ ,\ for\ BPF\ [\text{min}] \tag{1.2}$$

$$T_{cN} = N_t \cdot \sum_{k=1}^{n} T_{uk} + (N - N_t) \cdot \sum_{k=1}^{n} (T_{uk} - T_{uk+1})^+ ,\ for\ BTF\ [\text{min}] \tag{1.3}$$

where: n is the total number of operations in the manufacturing process; $(T_{uk} - T_{uk+1})^+$ is the positive difference of run times of consecutive operations k and $k+1$ (time difference is taken into account only if $T_{uk} > T_{uk+1}$); T_{un+1} is nil run time of an additional operation with index number $n+1$ that is added at the end of manufacturing process for calculation purpose in equation (1.3).

As the sum of differences of run times in equation (1.3) is smaller than the sum of run times of operations, it appears from equations (1.2) and (1.3) that the productivity ensured by BTF strategy is always higher than that of BPF strategy, whatever the size of transfer batch. Under these conditions, our research will further focus only on the batch production processes using the BTF strategy, in order to find a solution for increasing the productivity and robustness of their work schedules.

The productivity of a batch production schedule for a specific size of production batch (N) is generally given by the value of T_{cN}. However, in the case the production follows BTF strategy, the productivity of the schedule may be given also by the length of production cycles of transfer batches that compose the production batch. Let T_{cNtl} be the length of production cycle time for transfer batch with index number l during production.

The robustness of a batch production schedule is given by the size and distribution of operational slack times that are ensured for the work on transfer batches during the run of production process following BTF strategy. The larger the slack times provided for transfer batches in the production schedule, the higher the robustness of batch production process schedule will be. Moreover, knowing that at the end of the execution of the batch production process may generally appear problems induced by the wear of machines and tool, the workers fatigue and boredom states, etc., which lead to the extension of run times of operations for last transfer batches, the robustness of a batch production schedule is given by the size of operational slack times for last transfer batches. Therefore, the larger the slack times provided for the last transfer batches, the higher the robustness of a batch production schedule will be.

The productivity and robustness of a batch production schedule will be further investigated by taking into account only the operations time pattern in the production process. It is considered that a production process which needs a specific work quantity may be divided during the process planning stage in several operations with different operation time values.

2. Influence of operation times pattern on the productivity and robustness of production schedule

We assume that three operation time patterns may result after the planning stage of a multi-operational production process:

- a pattern with increasing values of the process operation run times (T_{uk});
- a pattern with decreasing values of the process operation run times (T_{uk});
- a pattern with random arrangement of process operation run time values (T_{uk}).

The following sequence of positive increasing numbers: $a_1, a_2, a_3, \ldots, a_n$, will be considered, where n is the number of operations in manufacturing process.

Because terms $N_t \cdot \sum_{k=1}^{n} T_{uk}$ and $(N - N_t)$ of equation (1.3) do not depend on the process operations run time pattern, the influence of the operation run times pattern will be studied only for the term $\sum_{k=1}^{n}(T_{uk} - T_{uk+1})^{+}$ in equation (1.3).

2.1. Influence on the productivity of the work schedule for entire production batch

2.1.1. Case I. Increasing values of the process operation run times

This hypothesis implies: $T_{u1} = a_1$, $T_{u2} = a_2$, …, $T_{un} = a_n$. Thus, the term $\sum_{k=1}^{n}(T_{uk} - T_{uk+1})^{+}$ value is as presented in equation (2.1).

$$\sum_{k=1}^{n}(T_{uk} - T_{uk+1})^{+} = \underbrace{(T_{u_1} - T_{u_2})}_{0} + \underbrace{(T_{u_2} - T_{u_3})}_{0} + \ldots + (T_{u_n} - \underbrace{T_{u_{n+1}}}_{0}) = T_{u_n} = a_n \ \text{[min]} \quad (2.1)$$

2.1.2. Case II. Decreasing values of the process operation run times

This hypothesis implies: $T_{u1} = a_n$, $T_{u2} = a_{n-1}$, …, $T_{un} = a_1$. Thus, the term $\sum_{k=1}^{n}(T_{uk} - T_{uk+1})^{+}$ value is as presented in equation (2.2).

$$\sum_{k=1}^{n}(T_{uk} - T_{uk+1})^{+} = \underbrace{(T_{u_1} - T_{u_2})}_{\geq 0} + \underbrace{(T_{u_2} - T_{u_3})}_{\geq 0} + \ldots + (T_{u_n} - \underbrace{T_{u_{n+1}}}_{0}) = T_{u_1} = a_n \ \text{[min]} \quad (2.2)$$

2.1.3. Case III. Random arrangement of process operation run time values

In order to obtain this random arrangement, the insertion method will be used. Two hypotheses will be taken into account, as follows:

- a random arrangement obtained by relocating a member of an increasing operation run times pattern: $a_1, a_2, a_3, \ldots, a_{k-1}, a_{k-j}, a_k, a_{k+1}, \ldots, a_{k+j}, \ldots, a_n$, where $2 \leq j \leq k-1$ and $a_{k-j} < a_{k-1}$. Thus, the term $\sum_{k=1}^{n}(T_{uk} - T_{uk+1})^{+}$ value is as presented in equation (2.3).

$$\sum_{k=1}^{n}(T_{uk} - T_{uk+1})^{+} = \underbrace{(a_1 - a_2)}_{0} + \ldots + \underbrace{(a_{k-1} - a_{k-j})}_{\geq 0} + \underbrace{(a_{k-j} - a_k)}_{0} + \ldots + (a_n - 0) = a_n + a_{k-1} - a_{k-j} > a_n \ \text{[min]} \quad (2.3)$$

- a random arrangement obtained by relocating a member of an decreasing operation run times pattern: $a_n, a_{n-1}, a_{n-2}, \ldots, a_k, a_{k-j}, a_{k-1}, a_{k-2}, \ldots, a_1$, where $2 \leq j \leq k-1$ and $a_{k-j} < a_{k-1}$. Thus, the term $\sum_{k=1}^{n}(T_{uk} - T_{uk+1})^{+}$ value is as presented in equation (2.4).

$$\begin{aligned}\sum_{k=1}^{n}(T_{uk} - T_{uk+1})^{+} &= \underbrace{(a_n - a_{n-1})}_{\geq 0} + \underbrace{(a_{n-1} - a_{n-2})}_{\geq 0} + \ldots + \underbrace{(a_k - a_{k-j})}_{\geq 0} + \underbrace{(a_{k-j} - a_{k-1})}_{0} + \\ &+ \underbrace{(a_{k-1} - a_{k-2})}_{\geq 0} + \ldots + \underbrace{(a_2 - a_1)}_{\geq 0} + (a_1 - 0) = a_n + a_{k-1} - a_{k-j} > a_n\end{aligned} \quad \text{[min]} \quad (2.4)$$

Analysing the above data, it can be observed that the batch production cycle time, T_{cN}, has the minimum value for increasing or decreasing operation times pattern, as can be seen in equation (2.5).

$$T_{cN}^{\min} = N_t \cdot \sum_{k=1}^{n} T_{uk} + (N - N_t) \cdot T_{uk}^{\max} \ \text{[min]} \quad (2.5)$$

where $T_{uk}^{\max}$ represents the longest run time of the production process operations.

As a consequence, a solution for enhancing the productivity of the work schedule of a multi-operational batch production process is to design the process with an increasing or decreasing operation times pattern.

2.2. Influence on the productivity of the work schedules for transfer batches

As follows, the influence of the increasing and decreasing values of the process operation run times on the transfer batches production cycle time will be checked. The transfer batch production cycle time is the time interval between the start of the batch manufacturing (in first process operation) and the end of the batch manufacturing (in last process operation).

In order to demonstrate the influence given by the operation run times pattern on the productivity of the work schedules for transfer batches, the following transfer batches will be considered: 1, 2, …, *l*, …, n_t, which are forming the manufacturing batch ($n_t = N/Nt$).

2.2.1. Case I. Increasing values of the process operation run times

This type of process operation run times pattern is exemplified in the Gantt chart presented in figure 1. The production cycle times for the first three transfer batches in the production batch are computed by using equations (2.6) ÷ (2.8).

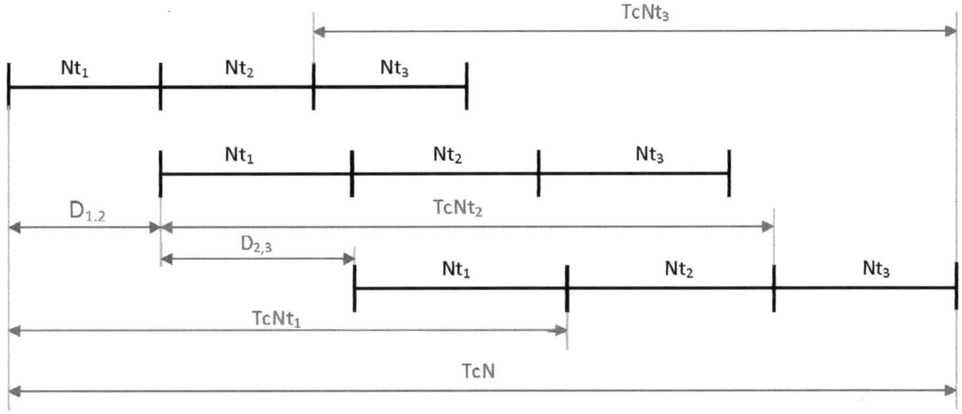

Figure 1. Schedule of transfer batches in the case of increasing values of the process operation run times, with indication of the production cycle times for batches.

$$T_{cNt1} = \sum_{k=1}^{n-1} D_{k,k-1} + N_t \cdot T_{u_n} = N_t \cdot \sum_{k=1}^{n-1} T_{uk} + N_t \cdot T_{u_n} = N_t \cdot \sum_{k=1}^{n} T_{uk} \text{ [min]} \quad (2.6)$$

$$T_{cNt2} = T_{cNt1} + N_t \cdot T_{u_n} - N_t \cdot T_{u1} > T_{cNt1} \text{ [min]} \quad (2.7)$$

$$T_{cNt3} = T_{cNt2} + N_t \cdot T_{u_n} - N_t \cdot T_{u1} = T_{cNt1} + (3-1) \cdot N_t \cdot (T_{u_n} - T_{u_1}) > T_{cNt2} \text{ [min]} \quad (2.8)$$

Based upon the above data, the following general equation can be used to calculate the production cycle time for the transfer batch with index number *l*:

$$T_{cNtl} = T_{cNt1} + (l-1) \cdot N_t \cdot (T_{u_n} - T_{u_1}) = N_t \cdot \sum_{k=1}^{n} T_{uk} + (l-1) \cdot N_t \cdot (T_{u_n} - T_{u_1}) \text{ [min]} \quad (2.9)$$

Thus, the transfer batches production cycle times ranking is: $T_{cNt1} < T_{cNt2} < \ldots < T_{cNtl} < \ldots < T_{cNtnt}$.

Hence, it results that the minimum transfer batch production cycle time can be calculated as shown in equation (2.10).

$$T_{cNt}^{\min} = N_t \cdot \sum_{k=1}^{n} T_{uk} \text{ [min]} \quad (2.10)$$

IOP Conf. Series: Materials Science and Engineering **682** (2019) 012018 doi:10.1088/1757-899X/682/1/012018

2.2.2. Case II. Decreasing values of the process operation run times

This type of process operation run times pattern is exemplified in the Gantt chart presented in figure 2. The production cycle times for the first three transfer batches in the production batch are computed by using equations (2.11) ÷ (2.13).

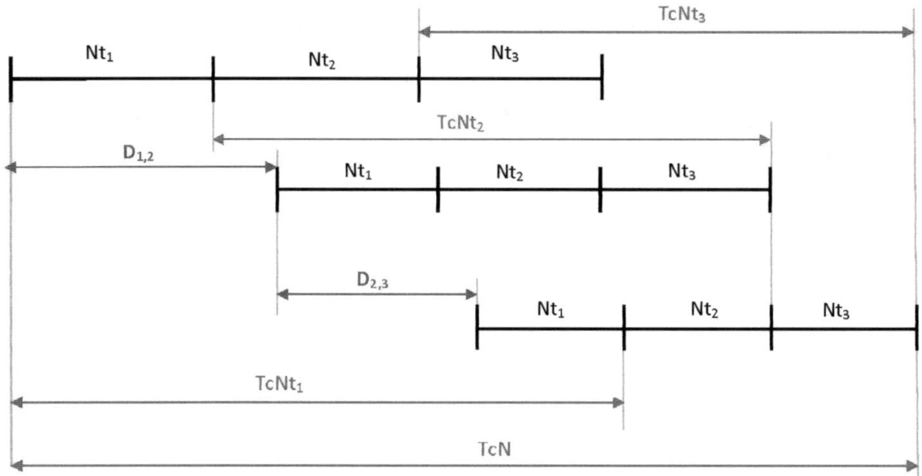

Figure 2. Schedule of transfer batches in the case of decreasing values of the process operation run times, with indication of the production cycle times for batches.

$$T_{cNt1} = \sum_{k=1}^{n-1} D_{k,k-1} + N_t \cdot T_{u_n} = N \cdot T_{u_1} - (N - N_t) \cdot T_{u_2} + N \cdot T_{u_2} - (N - N_t) \cdot T_{u_3} + \ldots + N \cdot T_{u_{n-1}} - (N - N_t) \cdot T_{u_n} + N_t \cdot T_{u_n} = N_t \cdot \sum_{k=1}^{n} T_{uk} + (N - N_t) \cdot (T_{u_1} - T_{u_n}) \quad [\text{min}] \qquad (2.11)$$

$$T_{cNt2} = T_{cNt1} + N_t \cdot T_{u_n} - N_t \cdot T_{u1} < T_{cNt1} \quad [\text{min}] \qquad (2.12)$$

$$T_{cNt3} = T_{cNt2} + N_t \cdot T_{u_n} - N_t \cdot T_{u1} = T_{cNt1} + (3-1) \cdot N_t \cdot (T_{u_n} - T_{u_1}) < T_{cNt2} \quad [\text{min}] \qquad (2.13)$$

Based upon the above data, the following general equation can be used to calculate the production cycle time for the transfer batch with index number *l*:

$$T_{cNtl} = T_{cNt1} + (l-1) \cdot N_t \cdot (T_{u_n} - T_{u_1}) = N_t \cdot \sum_{k=1}^{n} T_{uk} + (N - N_t) \cdot (T_{u_1} - T_{u_n}) + (l-1) \cdot N_t \cdot (T_{u_n} - T_{u_1}) \quad [\text{min}] \qquad (2.14)$$

For the last transfer batch, the cycle time is as follows:

$$T_{cNtn_t} = N_t \cdot \sum_{k=1}^{n} T_{uk} + (N - N_t) \cdot (T_{u_1} - T_{u_n}) + (N - N_t) \cdot (T_{u_n} - T_{u_1}) = N_t \cdot \sum_{k=1}^{n} T_{uk} \quad [\text{min}] \qquad (2.15)$$

Thus, the transfer batches production cycle times ranking is: $T_{cNt1} > T_{cNt2} > \ldots > T_{cNtl} > \ldots > T_{cNtnt}$.

Hence, it results that the minimum transfer batch production cycle time can be calculated as shown in equation (2.16).

$$T_{cNt}^{\min} = N_t \cdot \sum_{k=1}^{n} T_{uk} \quad [\text{min}] \qquad (2.16)$$

As can be seen in equations (2.9) and (2.14) and figures 1 and 2, the transfer batches have identical production cycle times, with the pattern of these times reversed for batches. In both cases, the minimum cycle time for production of a transfer batch was the same, see equations (2.10) and (2.16). In conclusion, the analysed patterns give the same productivity for the schedule of batch production process.

IOP Conf. Series: Materials Science and Engineering **682** (2019) 012018 doi:10.1088/1757-899X/682/1/012018

2.3. Influence on the robustness of the work schedules for transfer batches

The slack time of a work schedule for the transfer batch *l* at operation *k* represents the time interval between the end of the batch manufacturing at operation *k* and the beginning of the same batch manufacturing at operation *k+1*.

2.3.1. Case I. Increasing values of the process operation run times

This type of process work schedule is exemplified the Gantt chart presented in figure 3 and the calculation of the slack times of transfer batches in process operations is presented in equations (2.17) ÷ (2.19).

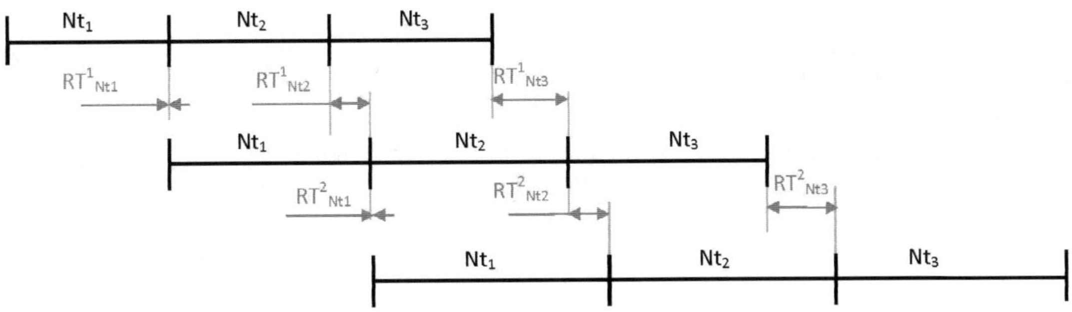

Figure 3. Schedule of transfer batches in the case of increasing values of the process operation run times, with indication of the slack times for transfer batches in operations.

$$RT^{k}_{Nt_1} = 0 \;\; [\text{min}] \tag{2.17}$$

$$RT^{k}_{Nt_2} = N_t \cdot (T_{uk+1} - T_{uk}) \;\; [\text{min}] \tag{2.18}$$

$$RT^{k}_{Nt_l} = N_t \cdot (l-1) \cdot (T_{uk+1} - T_{uk}) \;\; [\text{min}] \tag{2.19}$$

Thus, the transfer batches slack times ranking is: $RT^{k}_{Nt_1} \leq RT^{k}_{Nt_2} \leq RT^{k}_{Nt_l} \leq \ldots \leq RT^{k}_{Nt_{n_t}}$.

2.3.2. Case II. Decreasing values of the process operation run times

This type of process work schedule is exemplified the Gantt chart presented in figure 4 and the calculation of the slack times of transfer batches in process operations is presented in equations (2.20) ÷ (2.22).

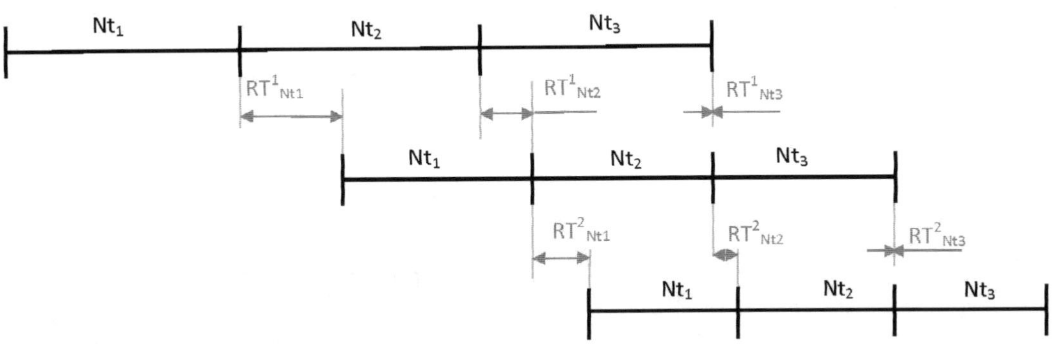

Figure 4. Schedule of transfer batches in the case of decreasing values of the process operation run times, with indication of the slack times for transfer batches in operations.

$$RT^{k}_{Ntn_t} = 0 \ \text{[min]} \tag{2.20}$$

$$RT^{k}_{Ntn_{t-1}} = N_t \cdot (T_{uk} - T_{uk+1}) \ \text{[min]} \tag{2.21}$$

$$RT^{k}_{Nt_l} = N_t \cdot (n_t - l) \cdot (T_{uk} - T_{uk+1}) \ \text{[min]} \tag{2.22}$$

Thus, the transfer batches slack times ranking is: $RT^{k}_{Nt_1} \geq RT^{k}_{Nt_2} \geq RT^{k}_{Nt_l} \geq ... \geq RT^{k}_{Nt_{n_t}}$.

As can be seen in equations (2.19) and (2.22) and figures 3 and 4, for both cases the transfer batches slack times are identical, with the pattern of these times reversed for the transfer batches.

However, an extension of the run time for the last transfer batch in the first operation, ΔT^{1}_{Nt3}, as shown in figures 5 and 6, has different consequences on the cycle times of both the transfer batch and production batch depending on the operation run times pattern.

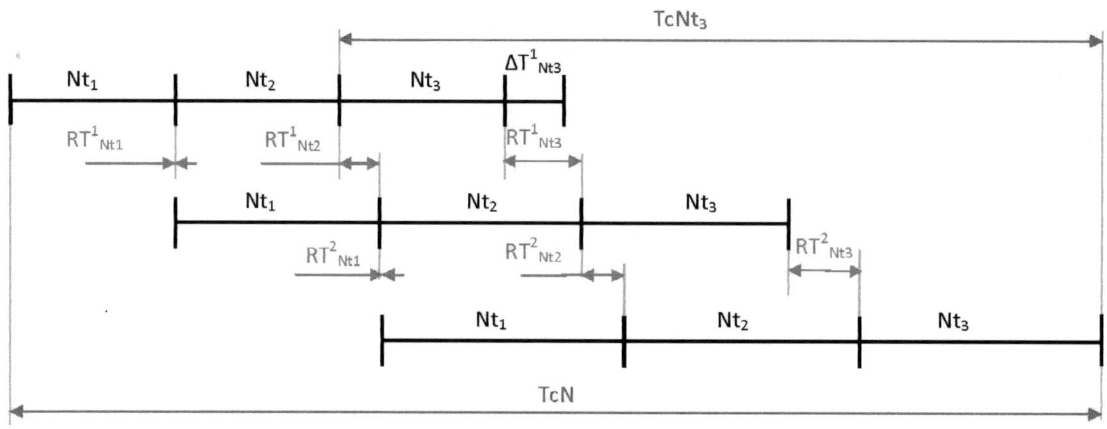

Figure 5. Schedule of transfer batches in the case of increasing values of the process operation run times, when an extension of the first operation occurs.

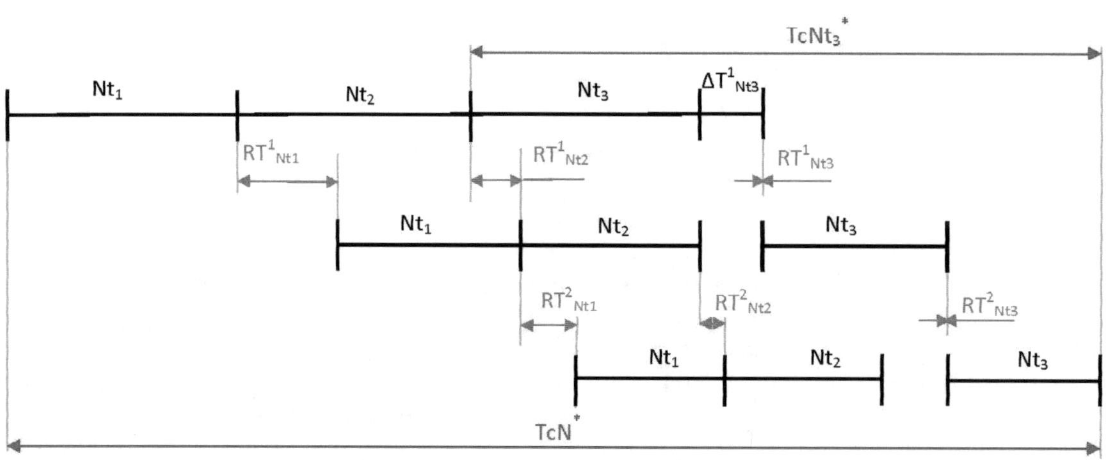

Figure 6. Schedule of transfer batches in the case of decreasing values of the process operation run times, when an extension of the first operation occurs.

In the case of increasing values of the process operation run times and when the extension of the run time for the last transfer batch in the first operation is smaller than the slack ($RT^1_{Nt3} > \Delta T^1_{Nt3}$), as seen in figure 5, the length of the cycles will not be modified for the last transfer batch and for the entire production batch as compared with those calculated with equations (2.9) and (1.3) and graphically presented in figure 1.

In the case of decreasing values of the process operation run times, any extension of the run time of the last transfer batch in first operation (ΔT^1_{Nt3}), as shown in figure 6, will extend the length of cycles for the last transfer batch and the entire production batch as compared with those calculated with equations (2.14) and (1.3) and graphically presented in figure 2.

As a consequence, a solution for enhancing the robustness of the work schedule of a multi-operational batch production process is to design the process with an increasing operation run times pattern because this pattern ensures larger slack times for the last transfer batches within the production batch.

3. Conclusions

This paper studied multi-operational batch production processes and the need to ensure high productivity and robustness for the work schedules of these processes. It was mathematically demonstrated that a favourable design of the processes, with increasing operation run times pattern, will minimize the production cycle time for the batch and will maximize the robustness of the work schedule of the process, in combination with the application of lot streaming strategy in the production flow.

4. References

[1] Catană M 2016 *Production and Operations Management*: *Course Notes* (Bucharest: POLITEHNICA PRESS)
[2] Neagu C, Niţu E, Melnic L and Catană M 2006 *Ingineria şi managementul producţiei: Bazele teoretice* (Bucharest: E.D.P. R.A.)
[3] Catană M 2010 *Proc. European Computing Conf. (Bucharest)* (WSEAS Press) p 82-87
[4] Baker K R and Trietsch D 2019 *Principles of Sequencing and Scheduling* (New Jersey: John Wiley & Sons)
[5] Cheng M, Mukherjee N J and Sarin S C 2013 *Int. J. Prod. Res.* **51** 7023–7046
[6] Dauzere-Peres S and Lasserre J 1997 *Operations Research* **45** 584–595
[7] Emmons H and Vairaktarakis G 2013 *Flow Shop Scheduling: Theoretical Results, Algorithms, and Applications* (New York: Springer Science + Business Media)
[8] Kalir A and Sarin S C 2001 *Decision Sciences* **32** 387–397
[9] Sarin S C and Jaiprakash P 2007 *Flow Shop Lot Streaming* (New York: Springer Science + Business Media)
[10] Trietsch D and Baker K R 1993 *Operations Research* **41** 1065–1076

IOP Conf. Series: Materials Science and Engineering **682** (2019) 012019 doi:10.1088/1757-899X/682/1/012019

Indoor path planning and obstacle avoidance simulation

T Savu and B A Jugravu

Manufacturing Engineering (TCM) Department, University POLITEHNICA of Bucharest, 313 Spl. Independentei, 060042 sector 6, Bucharest, Romania

savu@ctanm.pub.ro

Abstract. The paper is describing the work performed for choosing and designing a path planning and obstacle avoidance solution for a car-like vehicle moving in an indoor environment. The proposed algorithm first consists in providing the vehicle with a set of passing points generated by a Dijkstra path planning algorithm. The vehicle is using the points for establishing its path and is avoiding the obstacles using scanned data from a distance sensor. In a simulation environment, initial tests were performed for estimating correct values of the vehicle's forward and angular velocities and for estimating the effect of the scanning speed on the vehicle's behaviour. Obstacle avoidance tests were performed for identifying specific situations possible to appear due to high velocities or to low scanning speed. Even not always choosing a smooth avoidance path, the algorithm proved to find a way for avoiding the obstacle in a clear and fast enough manner.

1. Introduction

When designing autonomous ground vehicles, two topics are between those of main importance: the path planning and the obstacle avoidance capabilities. Path planning algorithms are treated at various levels of complexity, starting from simple line following procedures [1] to detailed analytical models [2]. Obstacle avoidance is also the subject of numerous works, from graduation thesis [3] to detailed contributions in exhaustive books [4].

Nevertheless, when designing an autonomous ground vehicle for a specific task, existing literature must be reviewed and knowledge to be applied must be selected as close as possible to the particular case. The scope of the work presented in this paper was to choose or to design a path planning and obstacle avoidance solution for a car-like vehicle moving in an indoor environment, more specifically in a building. Vehicle's hardware and software resources had to be considered not enough for performing high performance computing or for managing complex sensors sets. The paper is describing how a simulation environment was used for testing different options concerning the vehicle's parameters and for also testing an obstacle avoidance algorithm specifically designed for the above-mentioned purpose.

2. The path planning and obstacle avoidance algorithm

The proposed algorithm consists in the following:

- a set of compulsory points, to be used for moving from one building's space to a neighbouring one, are initially defined. When the vehicle must move from one point of the building to another, the set of the building's compulsory passing points is used by a Dijkstra path planning algorithm [5] for establishing the vehicle's path;

IOP Conf. Series: Materials Science and Engineering **682** (2019) 012019 doi:10.1088/1757-899X/682/1/012019

- the vehicle does not know the building's configuration, but knows its current position, is receiving the array of path points from the Dijkstra algorithm and is selecting the next point it must reach. After reaching one point, the vehicle is setting its target to the next one in the array of records, till the last point is reached;
- for moving toward the current target point, defined by a target vector based in vehicle's current position, the vehicle is first checking if there is any obstacle on the target vector's direction. If no obstacle is detected, the vehicle is moving forward;
- if an obstacle exists on the target vector's direction, then the vehicle is trying to avoid it by moving toward the obstacle's edge which is closer to the current moving direction.

For example, in figure 1, there is an obstacle on the target direction. The green lines indicate the obstacle edges. The vehicle will try to avoid the obstacle by starting to move to the rightmost green line, which is closer to the vehicle's speed vector.

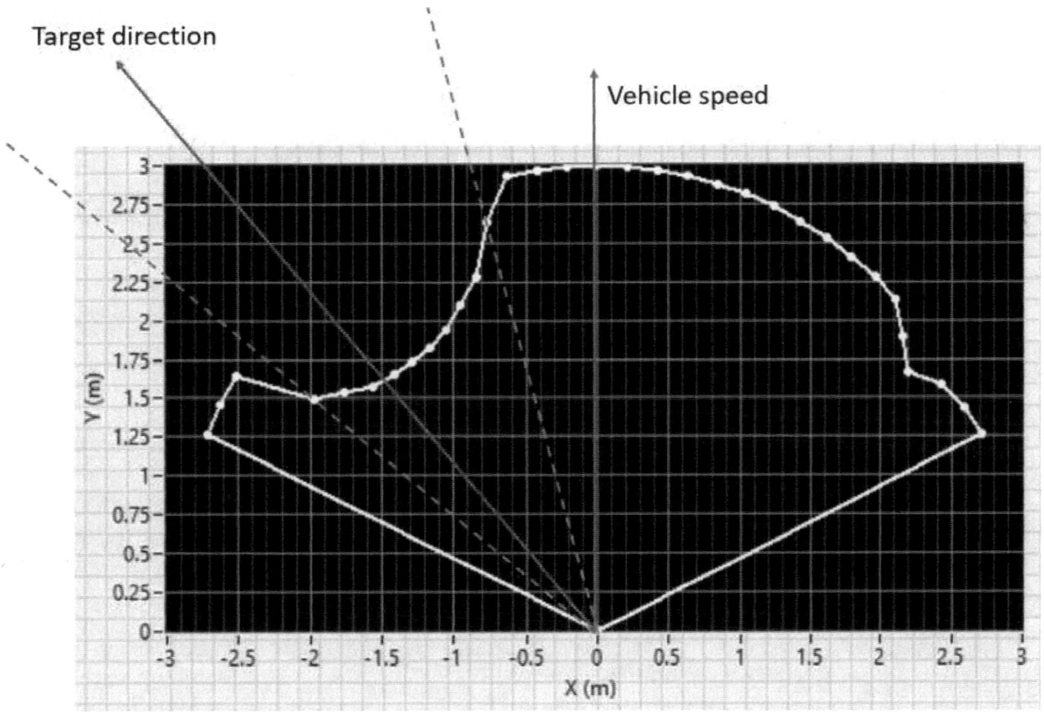

Figure 1. Radar points indicating an obstacle on the target direction.

3. Building the simulation environment

The simulation environment, which is reproducing part of the authors' faculty building, was initially built in a CAD environment (figure 2) and then described in the LabVIEW graphical programming environment [6], using the Robotics Starter Kit 1.0 functions library [7].

There are two areas in the environment which received special attention during the study:

- the departure room in the left side (figure 3), where the exit door is placed very close to a wall;
- an obstacle (figure 4), in the middle of one intermediate space, which is not known by the path planning software.

The building walls and the unknown obstacle were defined as box obstacles in the simulation environment (figure 5, figure 6). The data is stored in an array of clusters, where each cluster contains a subcluster with the three coordinates of the obstacle's mass center and two other values representing the obstacle's x-length and y-length. The obstacles' height was left at the default value of 1 m. All the obstacles were set to the "kinematic" state, making them to behave as if they have infinite mass.

IOP Conf. Series: Materials Science and Engineering **682** (2019) 012019 doi:10.1088/1757-899X/682/1/012019

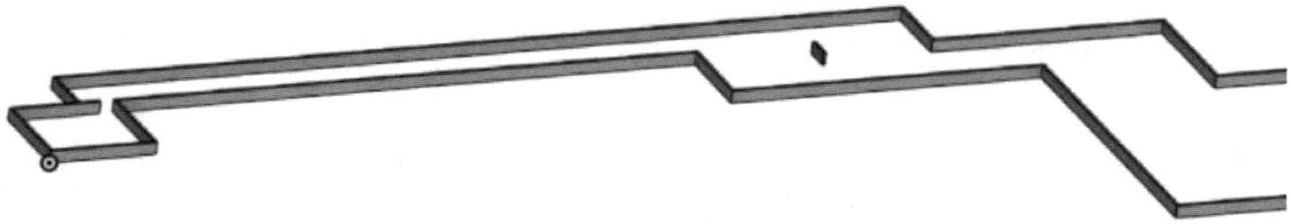

Figure 2. CAD model of the simulation environment.

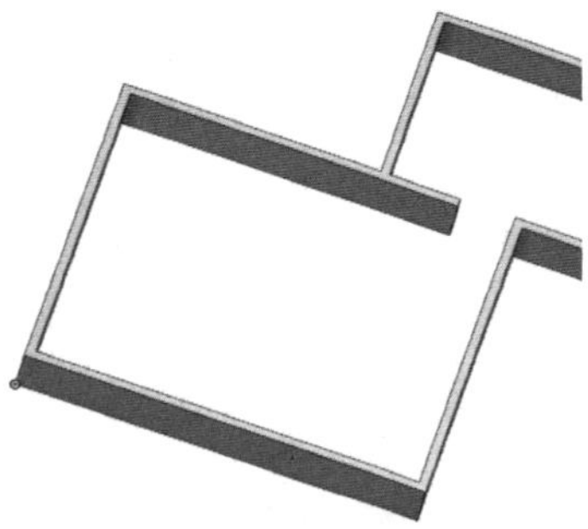

Figure 3. Detail of the departure room.

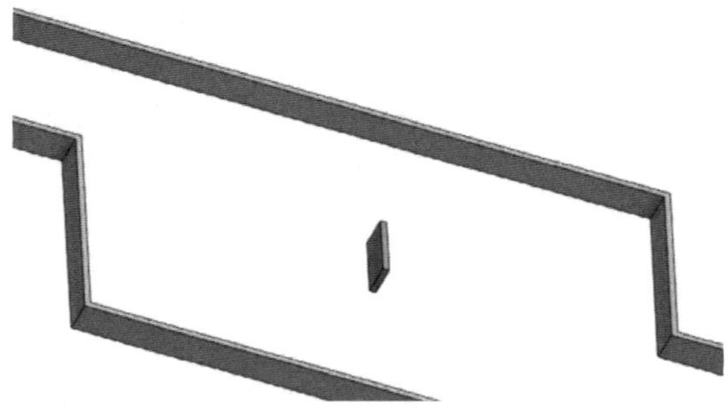

Figure 4. The obstacle in the intermediate space.

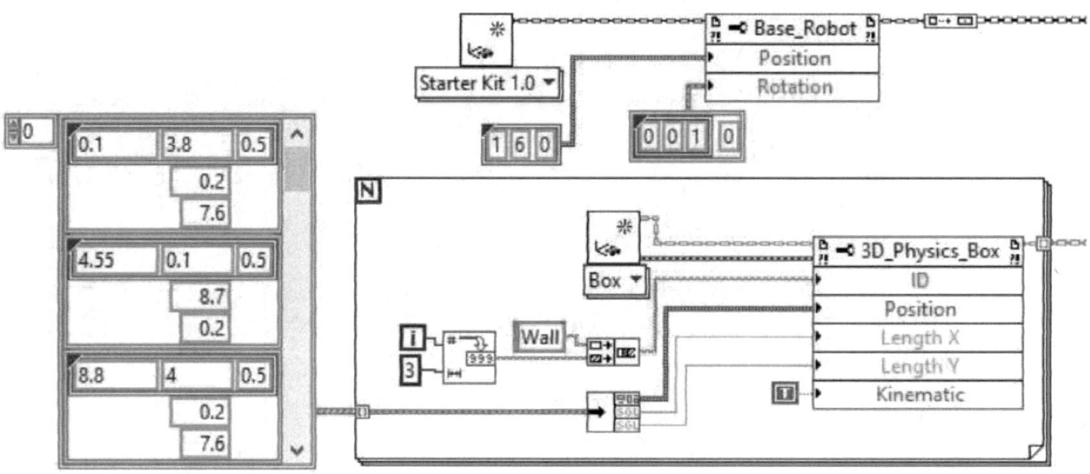

Figure 5. LabVIEW diagram for defining the simulation environment.

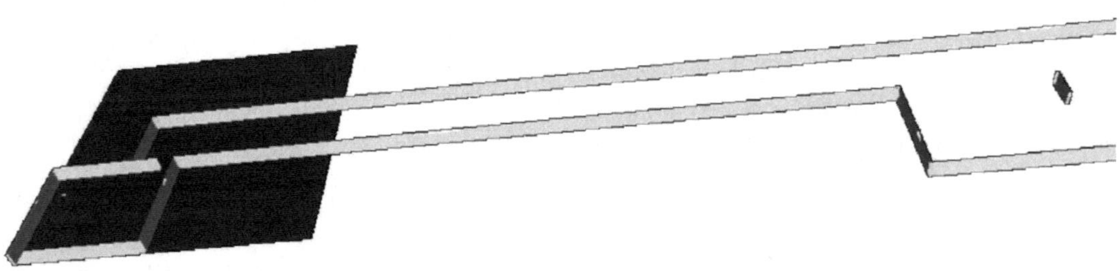

Figure 6. Building in the simulation environment.

The simulated situation is one when the vehicle must move from the left-side room to another space in the right part of the building. The points of the vehicle's path, generated by the Dijkstra algorithm, are stored in a global variable as an array of records, each record containing the x and y coordinates of one point (figure 7).

It is considered that a point on the path was reached by the vehicle when the distance to that point is less than a set value, initially 0.3 m.

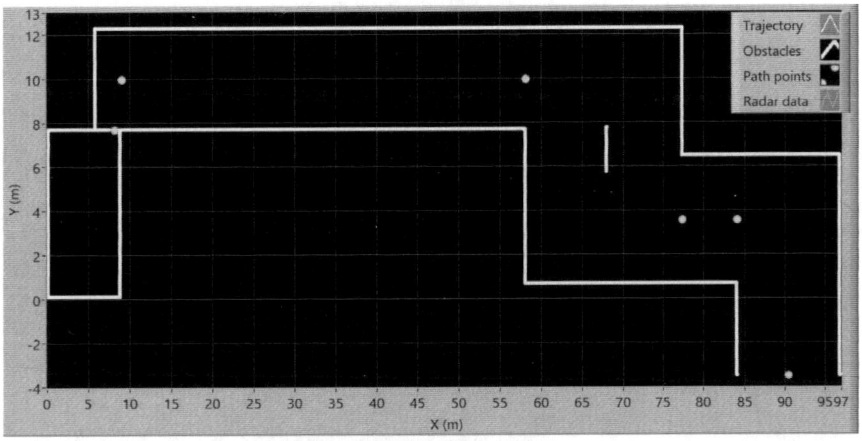

Figure 7. Points of the vehicle's path.

4. The simulated vehicle

Figure 8. The simulated vehicle.

The simulated vehicle (figure 8) is based on the one from the NI LabVIEW Robotics Starter Kit [8]. It has a steering frame composed of four simple fixed wheels with a radius of 50 mm. Only two wheels are active and the distance between them is 360 mm.

The active wheels are controlled, with a gear ratio of 2, by two 12V DC motors, with a maximum rotational speed of 152 rpm and a 2.1 N·m torque. Each motor has its own optical quadrature encoder with 400 pulses per revolution.

The vehicle is equipped with a PING))) 40 kHz ultrasonic distance sensor for distance measurements between 2 cm and 3 m. The sensor is rotated around the vertical axis using a servomotor. The rotation angle can vary between -65° and 65°. Angle increment of the scanned points was initially set at 4°.

The overall vehicle dimensions are 405 mm x 368 mm x 150 mm and the total weight is 3.6 kg.

The vehicle's initial position was set at coordinates (1,6,0) and its z-rotation was set to zero.

The vehicle's distance sensor is storing the measured data in two arrays, one containing the angles and the other the distances of the scanned points belonging to the detected obstacles. Both angles and distances are expressed in a polar coordinate system having its origin in the vehicle's origin.

According to the proposed algorithm, if the target vector's direction is outside the ±65° distance sensor's scanning interval, then the vehicle is moving toward that direction. If the target vector's direction is inside the interval but the scanned distance is at the 3 m limit, then there is no obstacle on the target vector's direction and the vehicle is also moving toward that direction.

If there is an obstacle on the target vector's direction at less than 3 m, then the start and end scanning angles between which the obstacle was detected are computed (figure 9) and the one closer to the target direction is selected as the updated vehicle direction.

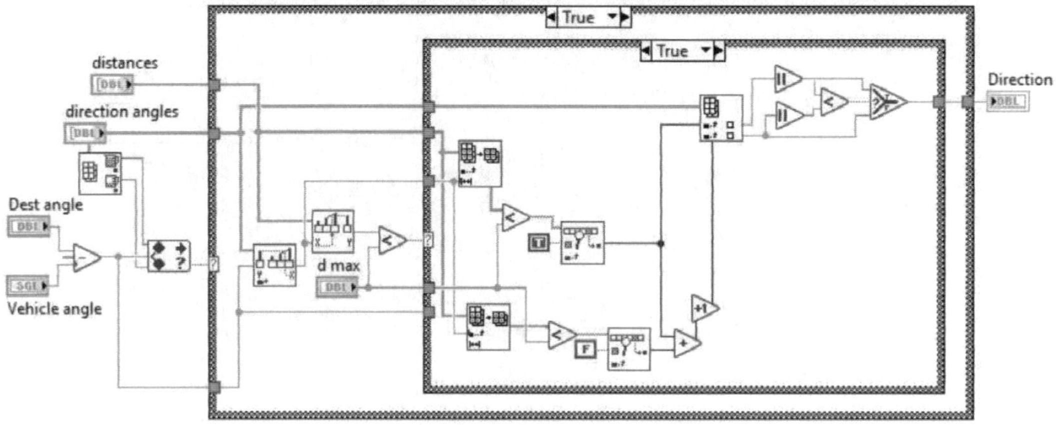

Figure 9. LabVIEW diagram for checking the presence of an obstacle and for computing the avoidance direction.

5. Initial tests

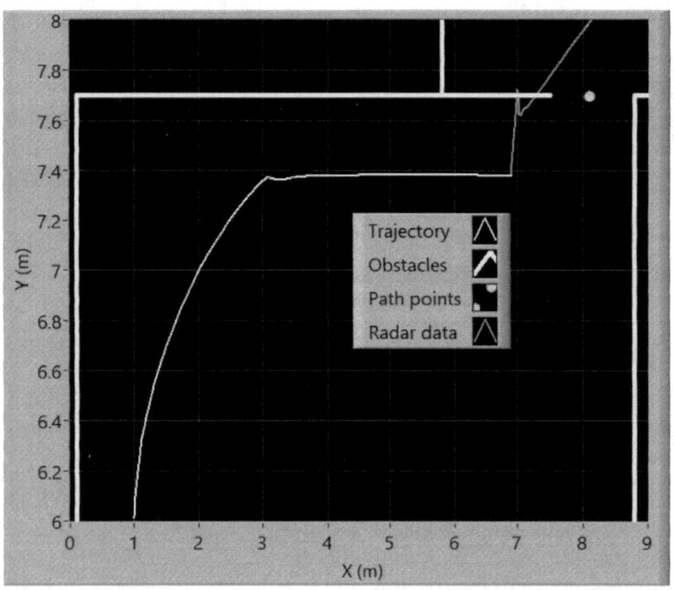

Figure 10. Moving along a wall due to big turning radius.

Initial tests were performed with the vehicle's maximum forward velocity set at 2 m/s and the maximum angular velocity set at 2 rad/s.

The tests showed that the angular velocity is too low. Trying to rotate toward the target point, the turning radius is big, the vehicle is reaching close to a wall and moving in contact with this (figure 10). The second conclusion is that, due to the small rotational speed of the distance sensor, the vehicle fails to collect enough scanned points till is reaching the exit door's vicinity.

When the maximum angular velocity was increased to 5 rad/s, the turning radius decreased, and the vehicle was able to position itself faster toward the target point (figure 11).

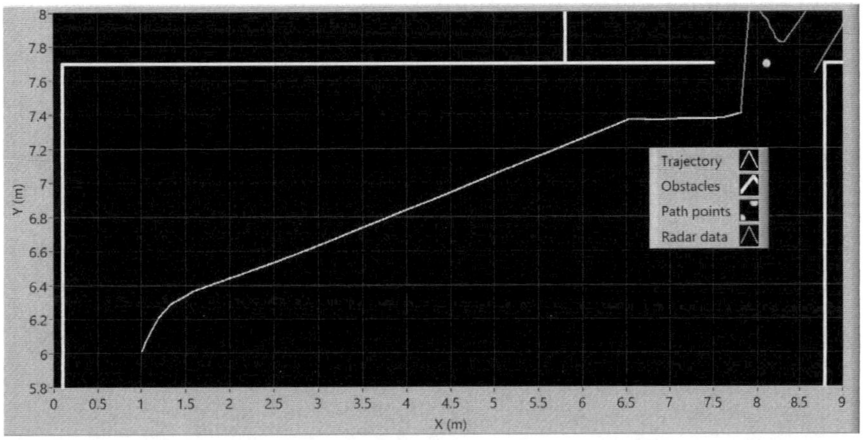

Figure 11. Faster positioning on the trajectory due to increased angular velocity.

Due to the increased rotational speed, there were tests during which the stability in moving toward the target's direction was initially lost, but finally the vehicle succeeded to position itself on the right direction (figure 12).

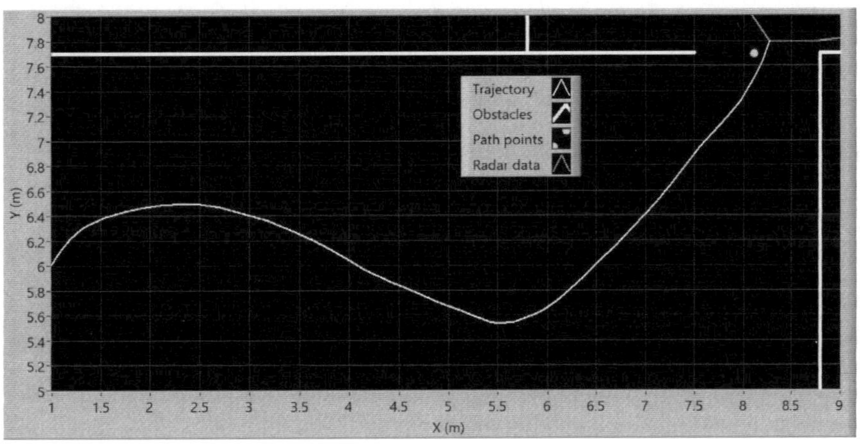

Figure 12. Stability loss due to increased rotational speed.

Figure 13. Vehicle stuck with backward wheels climbing on the wall.

Tests were performed with the maximum forward velocity set at 5 m/s and the maximum angular velocity set at 5 rad/s. It was only useful to demonstrate the simulator accuracy in reproducing the vehicle's dynamic, because it was reached a situation when the vehicle climbed a wall with the backward wheels and got stuck with the distance sensor scanning the floor (figure 13).

The maximum forward velocity of 2 m/s still proved too big, since in some situations the vehicle didn't succeed to break fast enough for almost hitting an wall (and was forced to move backward according to its "panic" algorithm) and also started turning right only after overcoming the second point (figure 14).

IOP Conf. Series: Materials Science and Engineering **682** (2019) 012019 doi:10.1088/1757-899X/682/1/012019

Reducing the maximum forward velocity to 1 m/s allows the vehicle to reach more precisely the specified path points. Some difficulties were still encountered when coming close to the wall on the right, because the distance sensor didn't have enough time to scan the entire interval, so the vehicle does not have the complete data about its surroundings (figure 15).

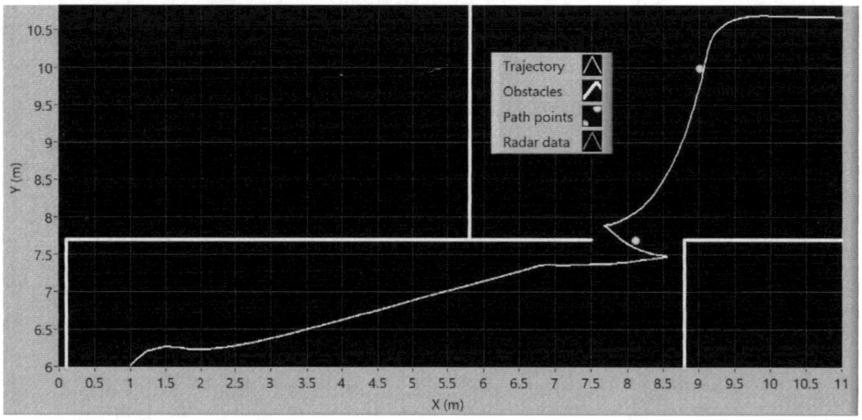

Figure 14. Moving backward in "panic" state and point overcoming due to high speed.

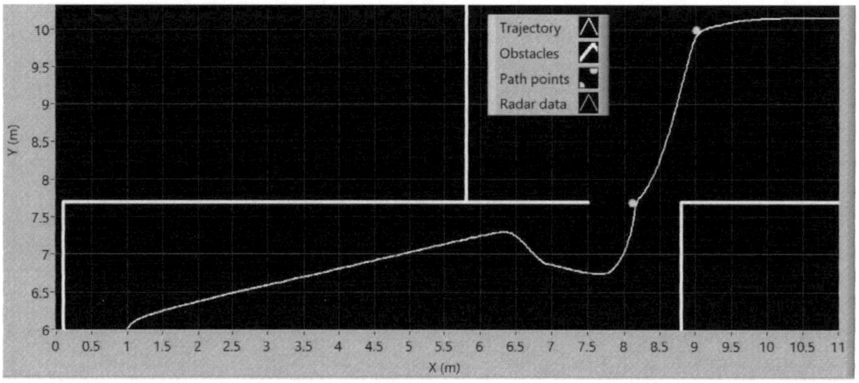

Figure 15. Difficulties in finding the door due to slow scanning speed.

Changing the starting point coordinates from (1,6, 0) to (1, 1, 0) gave enough time for the distance sensor to complete the scan, so the vehicle approached the room exit in a clean manner (figure 16).

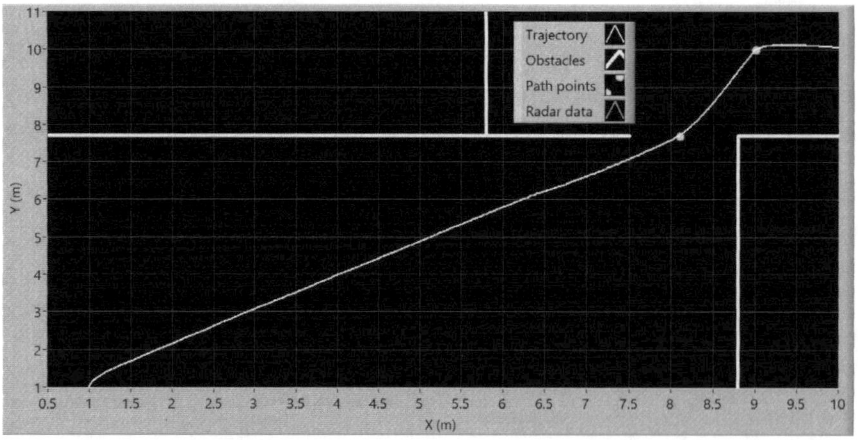

Figure 16. Clear path folloing when scanning data completed.

The radar data contains the information about the forward-left wall and also, when coming close to the exit door, the data allow to clearly identify the open space (figure 17, figure 18).

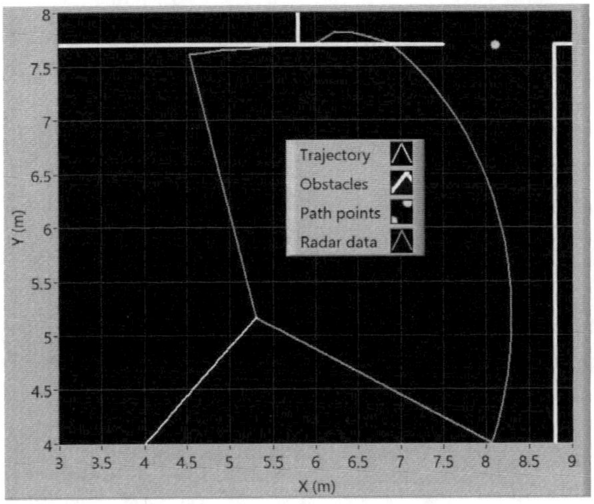

Figure 17. Forward-left wall radar data.

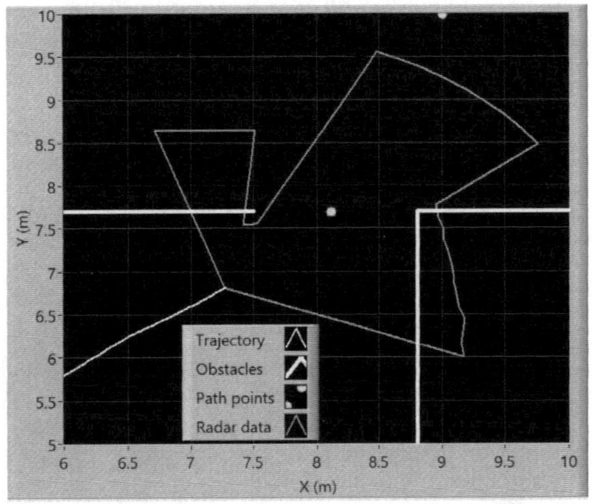

Figure 18. Open space (door) radar data.

6. Obstacle avoiding tests

When coming close to the unknown obstacle, due to the low scanning speed of the distance sensor, not all the obstacle's width is completely scanned at a certain moment. In figure 19 it can be seen that the scanning started from the left side of the obstacle (from the vehicle's point of view) and the obstacle's right side the vehicle is seeing a clear path, while in figure 20 the scanning started from the right side and there is a still unknown portion on the obstacle's left side.

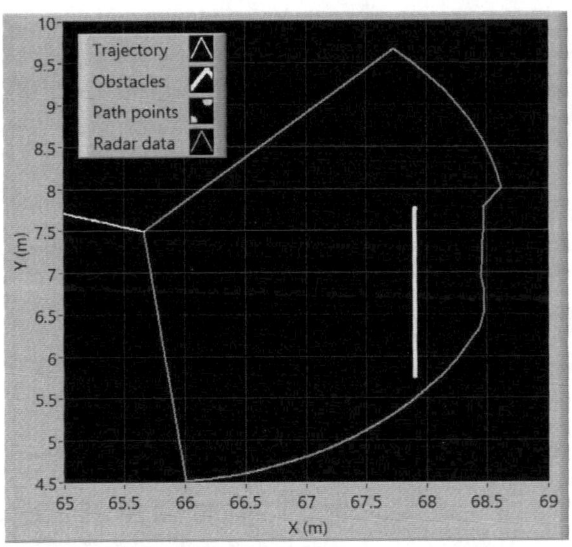

Figure 19. Obstacle radar data when scanning from the left.

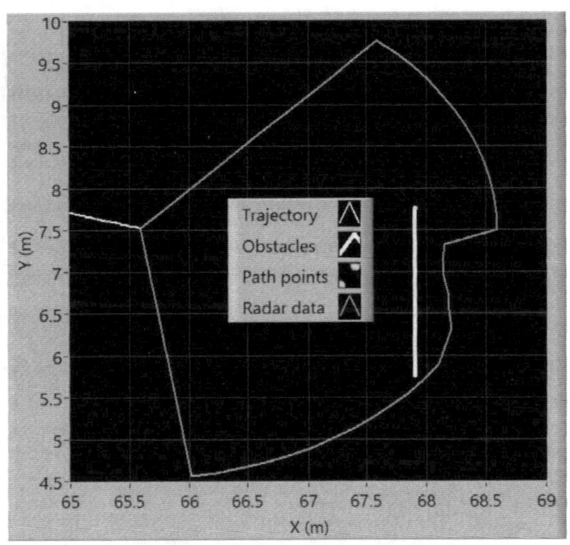

Figure 20. Obstacle radar data when scanning from the right.

In the first situation, the vehicle was choosing a detour point not right enough, so it reached close to the right side of the obstacle, close enough to hit it, entered the "panic" mode, moved back and then continued on a free path to the next target point (figure 21).

IOP Conf. Series: Materials Science and Engineering **682** (2019) 012019 doi:10.1088/1757-899X/682/1/012019

In a slightly different situation, the vehicle is sensing the right part of the obstacle from the beginning and it's moving on an avoiding trajectory. Because the distance sensor is scanning only in the ± 65° interval, the sensor is not detecting the obstacle when the vehicle is moving parallel to it. The vehicle started turning left too soon and the distance sensor detected now the right end of the obstacle. Even it was only an end of the obstacle, the vehicle entered the "panic" mode, moved backward and only afterwards continued to the next target point (figure 22).

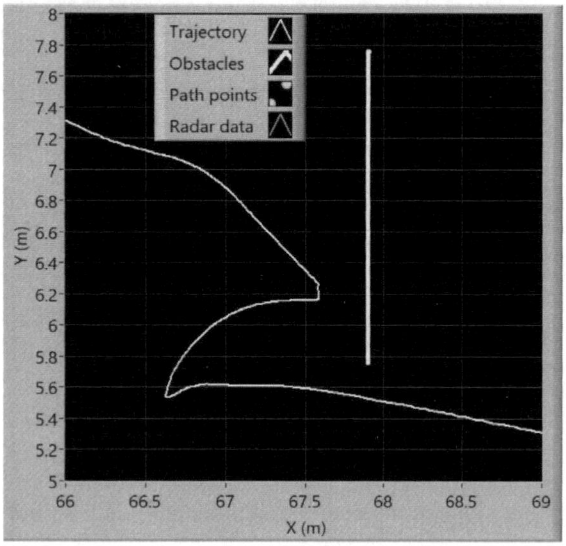

Figure 21. Incomplete obstacle avoidance when scanning from the left.

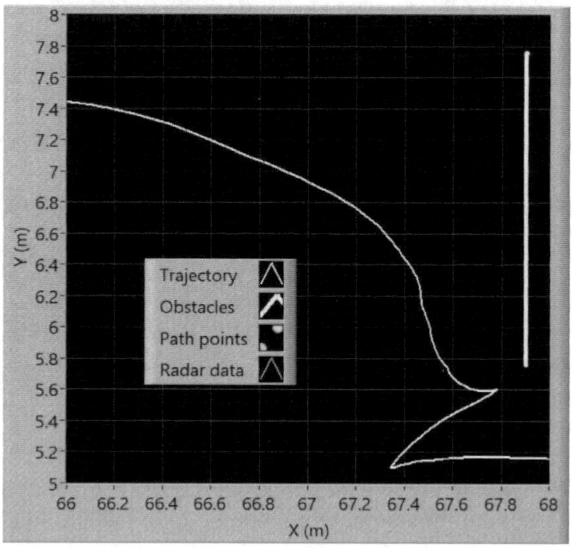

Figure 22. Obstacle re-appearance due to limited scanning interval.

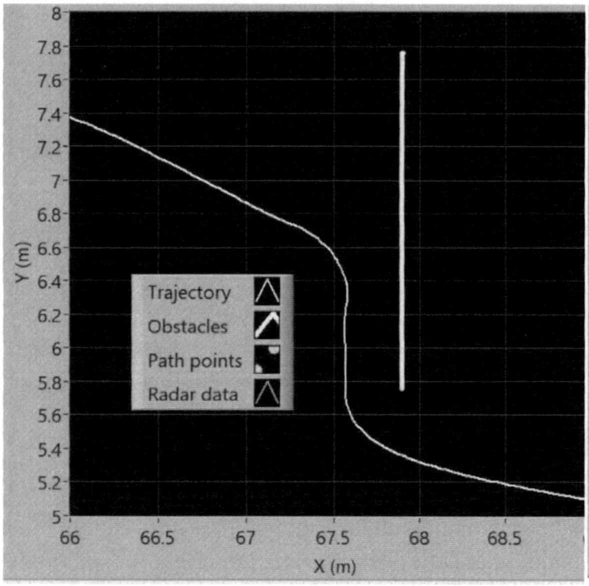

Figure 23. Smooth obstacle avoidance.

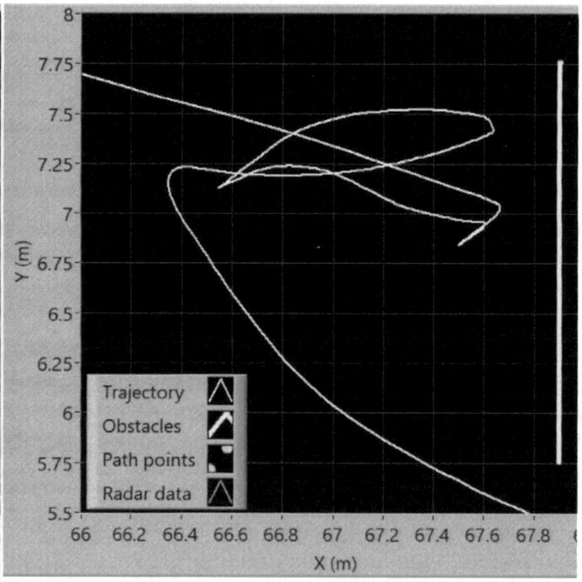

Figure 24. Difficulties in avoiding the obstacle.

An example of a situation when the obstacle avoidance run smoothly is presented in figure 23.

Final tests were performed with an increased forward velocity of 3 m/s. Due to the slow reaction, the vehicle had obvious difficulties in finding its way for avoiding the obstacle (figure 24) and in positioning on the exit door from the departure room (figure 25).

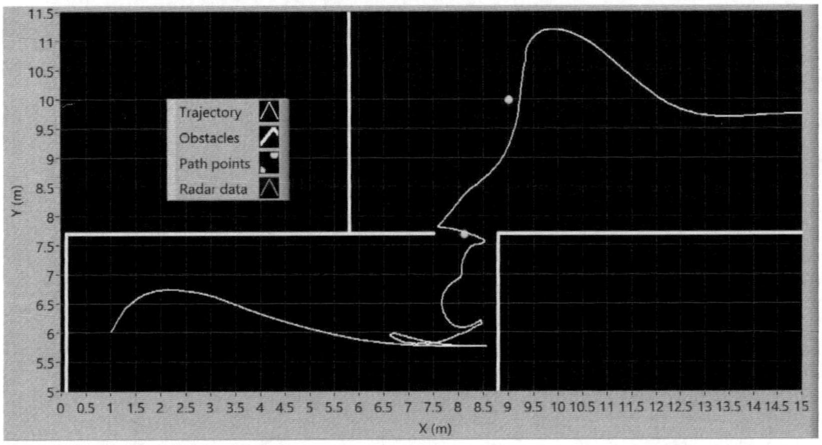

Figure 25. Difficulties in positioning on the exit door.

7. Conclusions and future developments

It can be concluded that, for the above described requirements, the chosen simulation environment allows an easy to use way of work for describing the simulated building, for bringing changes in its structure, for changing the vehicle's speed parameters, for changing the behavior algorithm and for obtaining relevant data.

Choosing a Dijkstra algorithm, run on a separate server, for computing the passing points in the building, proved to offer a simple way for the vehicle to determine its next steps without consuming high processing resources. Future studies will focus on changing the points data stored by the vehicle by including intermediate points needed for obstacle avoidance.

The simple obstacle avoidance algorithm proved to be simple and robust enough, allowing to identify the causes of various anomalies either in non-optimal vehicle speed parameters or in the low scanning speed of the distance sensor. Simulations will be designed for higher performance scanning sensors. Further studies will be performed for other vehicle configurations and dynamic parameters.

8. References

[1] Jibrail S F and Maharana R 2013 *PID Control of Line Followers* (Rourkela: National Institute of Technology)

[2] Fazlollahtabar H and Saidi-Mehrabad M 2015 *Autonomous Guided Vehicles – Methods and Models for Optimal Path Planning* (Cham: Springer International Publishing)

[3] Singh G 2015 *Sensor Fusion and Obstacle Avoidance for an Unmanned Ground Vehicle* (Daytona Beach: Embry-Riddle Aeronautical University)

[4] Fahimi F *Autonomous Robots – Modeling, Path Planning and Control* 2009 (New York: Springer Science+Business Media)

[5] Blokdyk G 2018 *Dijkstra's Algorithm: End-to-end Data Analysis* (Create Space Independent Publishing Platform)

[6] Bitter R, Mohiuddin T and Nawrocki M 2017 *LabVIEW: Advanced Programming Techniques, Second Edition* (Boca Raton: CRC Press – Taylor & Francis Group)

[7] Hossain A 2013 *Proc. Annual Conf. of American Society of Engineering Education* (Atlanta: ASEE Engineering Libraries Division) Paper ID #6116

[8] Siegwart R, Nourbakhsh I R and Scaramuzza D 2011 *Introduction to Autonomous Mobile Robots* (Cambridge: MIT Press)

IOP Conf. Series: Materials Science and Engineering **682** (2019) 012020 doi:10.1088/1757-899X/682/1/012020

Development of key performance indicators modules for small and medium-sized enterprises in production industry

A Tieber[1], D S Manolache[2] and M Gheorghe[2]

[1] PhD Student, University POLITEHNICA of Bucharest
[2] University POLITEHNICA of Bucharest, Manufacturing Engineering Department, 313 Splaiul Independentei, 060042 Bucharest, Romania

erfolg@alfredtieber.com, daniel.manolache@upb.ro , marian.gheorghe@upb.ro

Abstract. For small or medium-sized companies in the production industry, it is strongly advisable to intensify the use of key indicators in analysis and daily work, as well as open and honest communication throughout all levels of the organization as key factors for success. Although many enterprises now run customer relationship management systems to promote professionalization in the areas of customer satisfaction and customer service, key indicator systems are rarely used in a meaningful and optimal way. Massive potential for optimisation can be found in the whole system, so that a scientific survey study has been carried out in a number of companies from production industry. Based on the elements revealed by this study, a development of key performance indicators modules is proposed and synthetically described in the present paper. This development includes a series of key performance indicator – modules covering process and production management, sales management, customer and acquisition, human resources and innovation management, as an advanced structuring system in the field of key performance indicators.

1. Introduction

The demands placed on small and medium-sized enterprises, SMEs, are constantly increasing in the production industry. Order processing is becoming more complex, customer requirements and wishes are changing rapidly. This leads to SMEs taking care of most of the day-to-day operational business. Unfortunately, the further development and improvement of the own company and its business processes suffers [1]. In many cases, there is a lack of appropriate control mechanisms, or these are only partially implemented in the companies. This is a general problem of the production industry.

Key performance indicators, KPIs, are often used as a tool for monitoring, needs assessment and analysis. International standards such as DIN EN ISO 9001, ISO/IEC 27001, VDA 6. 2, ISO/TS16949 and others make explicit use of the possibilities of the targeted use of key figures [2]. However, the standards do not explain the individual KPIs in detail, but describe the use of KPIs to record, evaluate and analyze the specifications defined by the standards.

Accordingly, it is important that the use and introduction of key figures and KPI systems provide an accurate overview. It is necessary to analyse which are the company-specific expectations and then to verify which key figures and groups of key figures are relevant. Some examples of such key indicator groups are as follows.

For decades, financial ratios, the visualization of productivity and the economic assessment of the company have played a decisive role.

Since the 1980s, these monetary indicators have increasingly been supplemented by non-monetary indicators and developed into a management system that can depict the company in all its complexity and diversity [1]. For economic productivity, statements are made with respect to volume of sales by sales area or marginal return per customer, for instance. Economic efficiency can refer to the ratio between earnings and expenditure, performances and costs or also target costs and actual costs. Known profitability and financial key indicators are e.g. Cash flow, CF, Earnings before taxes, EBT, Earnings before interest and taxes EBIT, Earnings before interest, taxes and amortization EBITA, Profit or loss from ordinary business, EGT, Return on sales, ROS, Return on investments, ROI, Return on assets, ROA, Operating income from ordinary business, OOB [3].

Another large area is occupied by process key figures. These include production key figures, quality key figures, logistics key figures and supply chain key figures. The production capacity management, PCM, for example, is used to localise any potential cost optimisations and increase the efficiency of production processes. Material coverage period, MCP, considers the period of availability of stocked materials in a defined period. The key indicator is expressed in units of time (days, weeks or months). It analyses operating replenishment quantities. These include raw materials, utilities and supplies, as well as modules and components [3].

The quality rate examines the defect-free production quantity as a proportion of the total production quantity. The necessary data are available in the enterprise's internal documentation. The quality rate is an essential component of the quality management system (QMS). The rejection rate describes the portion of production that is defective and must therefore be discarded. A large number of process indicators have been developed over time and used more or less intensively. Some examples of these are machine hour rate, process cost rate and follow-up costs ratio [3].

Sales and marketing KPIs are very relevant for enterprises success. Mostly of these KPIs are known, but inconsequently applied Key sales indicators are used in the strategic and operational controlling of sales. The role of sales has changed fundamentally over recent decades. The task it once had as a communicative link between market needs and the internal business service offering is scarcely relevant today. An effective sales KPI system provides the enterprise with an information system for assessing all relevant sales, as well as the customer, competitor and market situation [4]. Key sales indicators that have a target-setting function offer the opportunity to manage the entire sales process. Examples for sales KPIs are purchasing power, sales success, turnover, contribution margin, sales cost structure and others [5].

The situation is similar with regard to key marketing figures. These are usually strongly linked to the sales process in their application. The use of marketing KPIs is limited across different sectors. Typically, the applied KPIs are also only known or communicated within the marketing pipeline. An international study stated, only around 48 % of surveyed companies have qualitative criteria for measuring the success, satisfaction or awareness of a campaign [6]. Typical marketing indicators are marketing productivity, early detection, brand strength or customer lifetime value.

Other interesting KPIs could be used in human resource and -development management. These include personnel costs ratio, recruitment costs, innovation rate, talent rate, number of graduates and others [3]. Unfortunately, however, many companies work inconsistently in these areas, the managers not being yet aware of the extraordinary potential that lies dormant in these two areas.

Product responsibles can develop proper KPIs based on user experience questionnaire (UEQ). This methodology should be designed and implemented in few steps. A UEQ KPI is calculated be a specific formula, and the results can be applied within the organisation [7].

In order to optimise the development of machine tools with efficient energy consuming, multiple process controls in the machine structure and a proper group of key performance indicators should be implemented. These are focusing on "inherent energy performance (IEP)". A structured method is proposed to accomplish "IEP indexes". A case study is performed confirming the utility of the considered method [8].

IOP Conf. Series: Materials Science and Engineering **682** (2019) 012020 doi:10.1088/1757-899X/682/1/012020

2. Development of KPI modules for SMEs in production industry

Taking into account the above elements, the development of a series of modules for the description of KPIs for SMEs in the production industry has been considered.

The term *module* describes a group of key indicators that includes a meaningful collection of individual KPIs that can be assigned to one or more business processes/ business areas. All modules together form a modular system. This enables the user to select an optimized number of key figures for his individual corporate management.

In order to develop an useful modelling system in industrial conditions, a scientific survey study has been designed for being addressed to a number of companies from production industry.

2.1. Survey study key facts

The designed scientific survey study was carried out in the DACH region (Germany, Austria and Switzerland/ German-speaking Europe) in the production industry, and some of its key figures are presented below.

An online questionnaire was developed and filled with 58 questions. A total of 102 companies were included in the study. Among other aspects, the companies were categorised based on the generated annual turnover, staff number and activity profile. So, 44.5 % of companies are those generating an annual turnover of more than EUR 100 million, 10.3 % - an annual turnover of 50 - 100 million EUR, 6.9 % - an annual turnover of 25 - 50 million EUR, 13.8 % - an annual turnover of 10 - 25 million EUR, and around 25 % - earned an annual turnover of less than 10 million EUR. Also, around 10 % of the companies participating in the study have an employment of more than 1 000 staff, 34.5 % of companies - an employment of 250 – 1 000 staff, and 34.5 % of companies - an employment of 30 – 50 staff, and 17.2 - 20.7 % - an employment less than 30 staff. The majority of survey participants (more than 59 %) were active in general management, and around 19,4 % of the participants were working in sales, marketing or purchasing department.

A distinction must be made between micro and small enterprises with an annual turnover up to 10 million EUR or a number of employees of up to 30 persons, medium-sized enterprises with an annual turnover of 10 - 100 million EUR or a number of employees of 30 - 500 persons, and large enterprises with an annual turnover of more than 100 million EUR or a number of employees of more than 500 persons.

2.2. Development of KPI - modules

A development of a main series of the KPI - modules, based on the elements of the considered study, has been proposed as follows.

2.2.1. KPI - module for process and production management

Process-, quality- and supply chain KPIs module is proposed as presented in Figure 1. It is essential that key indicators are automatically generated in this process. Looking at the production-specific key indicators for the process, quality and supply chain area, there is minimal leeway for suppliers/ sub-suppliers as this segment is particularly subject to a range of prescribed customer requirements. The study has confirmed that this is where customers specify the highest number of key indicators on the part of suppliers/ sub-suppliers. This is explained by the prescribed quality and documentation controls.

Significant elements associated to some of the *KPI-module for process and production management* components (Figure 1) are as presented in Table 1. In addition, the analytical representation of the *KPI-module for process and production management* is proposed as the system of the formulas (1) associated to the considered KPIs.

IOP Conf. Series: Materials Science and Engineering **682** (2019) 012020 doi:10.1088/1757-899X/682/1/012020

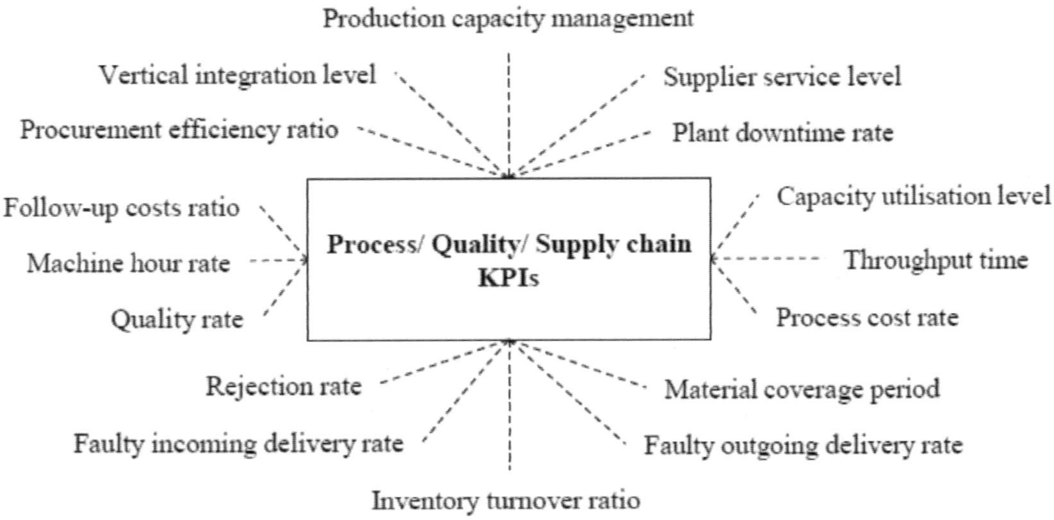

Figure 1. Process/ Quality/ Supply chain KPIs.

Table 1. Significant elements associated to some of the *KPI-module for process and production management* components

	Defining elements	Unit of measurement
FIDR:	Faulty incoming delivery rate	%
DFP:	Number of delivered faulty products	pcs or service-offers or .../ month or ...
DP:	Number of delivered products	
RR:	Rejection rate	%
RP:	Number of rejected pieces	pcs/shift or week or ...
PP:	Number of produced pieces	
...	...	...
ITR:	Inventory turnover ratio	%
RP:	Number of requirement products	pcs/week or month or ...
AIP:	Average inventory level of products	

$$\left\{ \begin{array}{l} KPI_1: FIDR = 100 * DFP/DP \\ KPI_2: RR = 100 * RP/PP \\ \dots \\ KPI_m: ITR = 100 * RP/AIP \end{array} \right\} \quad (1)$$

2.2.2. KPI - module for sales management

When considering the sales and market key indicators, it is clear that there is a strong connection between the company aims, the sales targets and hence also the associated key indicator requirements in the companies (Figure 2). For the controlling of sales among suppliers/sub-suppliers, however, a concentration on key indicators that analyse sales costs, time expenditure and internal sales expenses is also recommended.

Significant elements associated to some of the *KPI-module for sales management* components (Figure 2) are as presented in Table 2. In addition, the analytical representation of the *KPI-module for sales management* is proposed as the system of the formulas (2) associated to the considered KPIs.

Figure 2. Sales and Market KPIs.

Table 2. Significant elements associated to some of the *KPI-module for sales management* components

	Defining elements	Unit of measurement
SCS:	Sales cost structure	%
VSC:	Variable enterprise sales costs	euro/week or
TSC:	Total enterprise sales costs	year or …
MV:	Market volume	%
NT:	Enterprise net turnover	euro/week or
TNT:	Total realised net turnover	month or…
…	…	…
PPI:	Purchasing power index	%
PPR:	Purchasing power of a region	euro/
NPP:	National overall purchasing power	inhabitant-year

$$\left\{ \begin{array}{l} KPI_1: SCS = 100 * VSC/TSC \\ KPI_2: MV = 100 * NT/TNT \\ \dots \\ KPI_p: PPI = 100 * PPR/NPP \end{array} \right\} \quad (2)$$

2.2.3. KPI - module for customer and acquisition
Indicators for customer and acquisition processes (Figure 3) could prove to be suitable controlling mechanisms in order to increase the company's focus on customer needs and expectations.

Inefficiencies in the sales process and in assessing the market need to be removed from a standardised customer support mentality. Potential for optimisation can be found in the whole offer and negotiation process in sales.

Starting from the initial contact through support and processing in the offer stage, as well as negotiations, to the integration of the order in the company or customer handling in the event of a lost order, there are opportunities for increasing success. A targeted collection of offer and negotiation key indicators could reduce communication and comprehension problems in cooperation with the customer.

Figure 3. Customer and Acquisition KPIs

Significant elements associated to some of the *KPI-module for customer and acquisition* components (Figure 3) are as presented in Table 4. In addition, the analytical representation of the *KPIs-module for customer and acquisition* is proposed as the system of the formulas (3) associated to the considered KPIs.

Table 3. Significant elements associated to some of the *KPI-module for customer and acquisition* components

	Defining elements	Unit of measurement
CSR:	Contact success rate	%
ACC:	Arranged customer consultations	No./month or
TCC:	Total number of contacted customers	quarter or …
NCR:	New customer rate	%
NNC:	Number of new customers	No./month or
TNC:	Total number of customers	quarter or …
…	…	…
CI:	Contact intensity	%
NCC:	Number of contacts per customer	No./week or
TCC:	Total number of customer contacts	month or …

$$\left\{\begin{array}{l} KPI_1: CSR = 100 * ACC/TCC \\ KPI_2: NCR = 100 * NNC/TNC \\ \dots \\ KPI_q: CI = 100 * NCC/TCC \end{array}\right\} \quad (3)$$

2.2.4. KPI - module for human resources and innovation management

Human resource and innovation management KPIs module is proposed as presented in Figure 4. Recruiting, employee support and development, staff training, employee satisfaction and much more fall under the remit of HR management.

Significant elements associated to some of the *KPI-module for human resources and innovation management* components (Figure 4) are as presented in Table 4. In addition, the analytical representation of the *KPIs-module for human resources and innovation management* is proposed as the system of the formulas (4) associated to the considered KPIs.

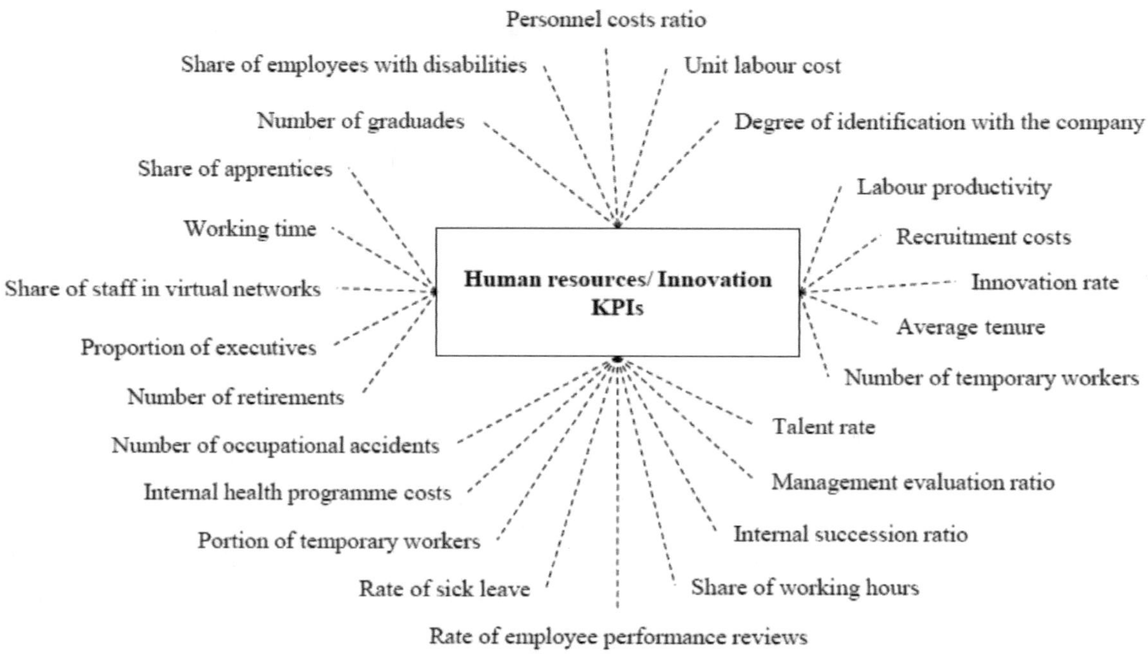

Figure 4. Human resources/ Innovation KPIs.

Table 4. Significant elements associated to some of the *KPI-module for human resources and innovation management* components

	Defining elements	Unit of measurement
RSL:	Rate of sick leave	%
SLT: WT:	Total of sick leave time Total working time	hrs/week or month or year
PTW:	Portion of temporary workers	%
NTW: TS:	Number of temporary workers Total staff	No./week or month or …
…	…	…
REPR:	Rate of employee performances reviews	%
NEPR: TSN:	Number of employee performance reviews Total staff number of the department (division, enterprise, etc.)	No./week or month or quarter or year

$$\begin{cases} KPI_1: RSL = 100 * SLT/WT \\ KPI_2: PTW = 100 * NTW/TS \\ \dots \\ KPI_r: REPR = 100 * NEPR/TSN \end{cases} \quad (4)$$

The use of HR key indicators would enable companies to set new incentives in times of skill shortages and a high job change rate among employees within industries, in order to distinguish themselves from the market. The more transparently and objectively the value as well as performance of an employee can be analysed and assessed, the more accurately potentials can be leveraged and individual strengths localized.

The KPI - modules presented above together with other developed modules - associated to economic KPIs, profitability and financial KPIs, offers-price-delivery KPIs and marketing KPIs - makes it possible to generate a lean and efficient key figure model based on knowledge of the individual company situation.

3. Conclusions

The production, sales, personnel expenditure, employee quality and innovation performance of the company can be depicted in a structured manner.

The series of the proposed KPI - modules for process and production management, sales management, customer and acquisition, human resources and innovation management represent an advanced structuring system in the field of key performance indicators.

Intelligent KPI management in human resources can tremendously increase the likelihood that people decide to join the company and stay there for the medium to long-term. Transparency in analysis and daily work, as well as open and honest communication throughout all levels of the organisation constitute key factors for success in this endeavour.

Appropriate use of digital information platforms will become incredibly important and decisive in the future. For success as a small or medium-sized company in the production industry, it is strongly advisable to intensify the use of key indicators in innovation and human resource management. In turn, this means that managers in particular have to bear the responsibility to introduce and implement this philosophy on a consistent basis. Communication and continuous work on the relevant processes are therefore imperative.

4. References

[1] Pleier N 2008 *Performance-Measurement-Systeme und der Faktor Mensch: Leistungssteuerung effektiver gestalten* (Wiesbaden: GWV Fachverlage GmbH)

[2] Brückner C 2009 *Qualitätsmanagement für die Automobilindustrie – Grundlagen, Normen, Methoden* (Düsseldorf: Symposion Publishing GmbH)

[3] Krause H-U and Dayanand A 2008 *Controlling-Kennzahlen: zweisprachiges Handbuch Deutsch/ Englisch* (Munich: Oldenbourg Verlag)

[4] Dietzel A 2013 *Vertriebscontrolling optimieren: Grundlagen und Praxis* (Wiesbaden: Gabler Verlag)

[5] Ossola-Haring C 2016 *11 Irrtümer über Kennzahlen: Mit den richtigen Zahlen führen* (Wiesbaden Springer Fachmedien,)

[6] Pufahl M and Janning R 2016 Customer Journey Management als integrierte Methodik zum kundenzentrierten Geschäftsmodell–Marktstudie 2016 (https://www.ec4u.com/cjm-studie-2016/)

[7] Hinderks A, Schrepp M, Mayo F J D, Escalona M J and Thomaschewski J 2019 Developing a UX KPI based on the User Experience Questionnaire *Computer Standards & Interfaces* **65** 38 – 44

[8] Tuo JB, Liu F and Liu PJ 2019 Key performance indicators for assessing inherent energy performance of machine tools in industries *Int'l J. of Production Research* **57** 6 1811-24

AUTHOR INDEX

Institute of Physics
Dirac House, Temple Back
Bristol BS1 6BE UK

ISSN: 1757-8981
ISBN 978-1-7138-0934-0